HUMAN NATURE EXPLORED

Books by GEORGE MANDLER

The Language of Psychology (with William Kessen)
Thinking: From Association to Gestalt (with Jean Matter Mandler)
Mind and Emotion
Mind and Body: Psychology of Emotion and Stress
Cognitive Psychology: An Essay in Cognitive Science

Human Nature
Explored

GEORGE MANDLER

University of California, San Diego
University College London

New York Oxford
Oxford University Press
1997

Oxford University Press

Oxford New York Athens Auckland Bangkok Bogota Bombay Buenos Aires
Calcutta Cape Town Dar es Salaam Delhi Florence Hong Kong
Istanbul Karachi Kuala Lumpur Madras Madrid Melbourne
Mexico City Nairobi Paris Singapore Taipei Tokyo Toronto Warsaw

and associated companies in
Berlin Ibadan

Published by Oxford University Press, Inc.
198 Madison Avenue, New York, New York 10016

Oxford is a registered trademark of Oxford University Press

Library of Congress Cataloging-in-Publication Data

Mandler, George.
Human nature explored / George Mandler.
p. cm.
Includes bibliographical references and indexes.
ISBN 0-19-511223-7
1. Psychology. 2. Human behavior. I. Title.
BF121.M363 1997
150—dc21   97-22045

1 3 5 7 9 8 6 4 2

Printed in the United States of America
on acid-free paper

To the memory of Suse
and all my cousins who perished

Preface

As a great believer in the over- and multidetermined nature of human thought and action, I find a number of reasons that led me to write this book:

A commitment to the notion that complex human behavior can be explained in terms of basic psychological processes.

A combination of this commitment with a desire to counterbalance the current preoccupation with developing—often unsubstantiated—stories about the genetic basis of complex social behavior. I have always had a distaste for simpleminded invocations of genetic mechanisms to explain human behavior, in particular when such invocation is done without any intensive attempt to understand the behavior in psychological and social terms. At the same time, I am committed to understand the basic evolutionary processes that have shaped human behavior.

Finally, I thought a retrospective view of my work over the past several decades, informed by a particular view of the human mind, would be desirable to have in one place.

Since the mid-1970s I have divided my professional activities between a continuing career as an experimental psychologist, particularly in the area of memory, and a more contemplative streak that took on such issues as emotion and consciousness. Some of these activities found a place in books, some as chapters in collections written and targeted for a variety of audiences. Wherever they were, however, I saw these pieces as part of a continuing and coherent effort to understand important aspects of human psychology in an evolutionary context. This book tries to put it all together. Given the history of my interests in these areas, some of the chapters are revised versions of previous material, and this is particularly true of chapters 4, 5, 6, 7, 10, the Appendix, and parts of chapter 11; conversely, chapters 1, 2, 3, 8, 9 and parts of chapter 10 and 11 are new.

Writing this book was a transcontinental project, in particular after 1989 when I started to split the year between London and La Jolla, California. My dual appointment at the University of California, San Diego, and at University College London made the facilities of both these institutions available to me, and I am grateful to both universities for supporting my efforts.

Many friends influenced my thinking about many of the issues raised here, and I want to acknowledge my colleagues in San Diego and London who helped me think about some of these topics. A particular role has been played by my wife, Jean Matter Mandler, and my sons, Peter and Michael, who read much of the material included here and tried to keep me from making too many blunders. I am most grateful to Ellen Berscheid, Alan Fridlund, and Philip Kitcher, who read and commented on most or all of an earlier draft and helped me avoid errors and incorporate many new thoughts and organizations. Kimberly Jameson and Patricia Kitcher read and helped with selected sections and I am indebted to them. Most of the material in this book was the basis of my lectures in a course on explorations in human nature, which I gave to an always-interactive and helpful audience of University of California, San Diego, seniors between 1987 and 1994. Finally, I want to thank Joan Bossert, my insightful and always helpful editor, and Jon Lomberg for his invaluable help in designing the cover.

La Jolla and London   G.M.
September 1996

Contents

ONE   Introduction: Themes, Tasks, and Methods   3

TWO   Historical and Biological Constraints   9
        Myths and Human Thought about Human Beings   9
        Biological Constraints   15

THREE   Evolution and the Genetics of Behavior   21
        General Principles of Evolution   21
        The Evolution of Humans   28
        Sociobiology and Behavior Genetics   32

FOUR   Minds, Bodies, and Schemas   40
        Minds and Bodies   42
        The Nature of Schemas   43

FIVE   Consciousness   46
        A Constructivist Approach to Consciousness   48
        The Functions of Consciousness   49
        Limited Capacity and the Utility of Consciousness as a Serial Device   52

The Feedback Function of Consciousness   54

Constructions of Consciousness   56

Memory and Consciousness   60

Some Speculations about the Evolution of Consciousness   61

SIX   Emotion   66

Emotions: Approaches, Problems, and Myths   66

Understanding Hot Emotions: Causes, Consequences,
    and Constructions   69

Difference Detection: Discrepancies Are Important   70

The Construction of Emotion   71

Some Special Problems of the Emotions   74

Evaluation: The Culture of Emotion   76

The Contemporary Scene   78

Some Speculations about the Evolution of Emotion   82

SEVEN   Values in Human Thought and Action: Origins and Functions   85

What Is a Value?   88

Sources of Values   90

Conclusion: Interdependence, Biology, and a Contradiction   98

EIGHT   The Social Fabric: Aggression and Other
              Human Characteristics   100

The Functions of Society   100

Aggression   101

A Look at Alternative Social Organizations   108

Social Alternatives: Aggression and Cooperation   110

NINE   Differences Among People: Intelligence and Gender   112

The Problem of Intelligence   112

Sex and Sex Differences   118

Some General Comments on Individual Differences   121

TEN   Morality, Freedom, and Power   123

Morality and Human Nature   123

Rationality and Universality   126

Freedom, Constraints, and Power: What Does It Mean to Feel Free?   127

ELEVEN   Cognition and Language   137

Categories of Thought   138

Language 139

World and Mind: Structure and Representation 141

Reductionism: Mind Is not Brain, or Vice Versa 143

Concluding Thoughts 144

APPENDIX Psychology as a Reflection of Cultural Values 147

Prologue 147

Introduction 147

Choicepoints in the History of Psychology 149

Physics Envy and Doing Psychology without Experiments 166

Envoi 168

NOTES 169

REFERENCES 185

INDEX OF NAMES 199

INDEX OF SUBJECTS 205

HUMAN NATURE EXPLORED

ONE

Introduction

Themes, Tasks, and Methods

IN DISCUSSING HUMAN nature I approach the topic primarily from a psychological point of view. I concentrate on those characteristics that are typical of individual human thought and action. I also try to understand what possible evolutionary factors of natural selection may have affected the genetic underpinning of these behaviors. On the other hand, I am equally interested in how social situations modify basic human characteristics. The possible range of a topic like "human nature" is vast indeed, and I could not possibly cover all instances that have been enumerated in the past. In addition, there are a number of issues about human nature that I have avoided, either because they are well beyond my competence or because they have been adequately treated elsewhere.[1] I leave it to the reader to explore those aspects I may have missed.

If there is a single theme that applies to most if not all of the psychological material I present, it is the attempt to find some few basic components of human mental life that can account for more complex phenomena. For example, I depend to a large extent on schema theory to understand our interactions with the world, and in particular the construction of human values. Similarly, a central part is played in my story by the notion of discrepancy or interruption, the postulation that one of the important products of (at least) mammalian evolution is the detection of differences between the current and previous states of the world.

That preference for deriving complex phenomena from more basic, if not necessarily simpler, structures also informs my approach to most evolutionary,

3

genetic explanations of human, and particularly social, phenomena. Some of the current attempts to find genetic bases for complex social phenomena demonstrate a lack of enterprise and a theoretical sluggishness typical of much of contemporary social science. Since genetic explanation of behavioral phenomena are, at least for complex human behavior, usually conjectures and speculations without any evidentiary basis, it is much easier to invoke such an explanation than to attempt a detailed and extensive analysis of the phenomenon involved before giving up and assigning it to genetics. Surely, there is no disagreement that our evolutionary history is a necessary condition for everything that we think and do. The question is what kinds of building blocks—primarily the products of genetic factors—do we wish to invoke for our explanations and conjectures. I prefer to see relatively broad, widely accepted, and foundational aspects of behavior as being primarily the ones on which we rely for a genetic account.

This book is not intended to be a review of the literature on human nature or even of the relevant psychological literature. It is centered on themes I have considered in the past, and in that context it does take into account the available evidence and other contributions relevant to the arguments I make. Finally, when selecting basic characteristics, I have tried to speak about aspects of human thought and action that are relatively culture free. Thus consciousness (but not its contents) should operate the same way for Indians, Eskimos, and Yuppies; the same mechanisms of emotional experience (but not necessarily the functions and contents of emotion) are posited to apply to Scots, Kwa Zulus, and Aztecs. This leaves out many of the myths and stories that view human nature as social constructions, and it also leaves out some mainstream speculations such as Freud's that are wholly or partially culture bound. I do, however, address problems of human variation within their social and cultural contexts.

While I cannot cover all of the characteristics that have ever been considered part of human nature, I do explore some of them, such as aggression, in order to suggest their variability and possible underlying processes. For other features, I explore the extent to which they may be products of our history, not our nature. I am therefore concerned in part in developing the context—historical and social—that came to make us believe certain things about ourselves. The main task I have set for this volume is to give some appreciation of the basic constituents of what it means to be human. I try to answer, though surely only partially, such questions as: What is human nature? What do we mean by human nature? What is it that we refer to in such phrases as "It is only human nature," "You cannot change human nature," or "To err is human"? The reference in such phrases seems to be to the bedrock aspects of human characteristics, to those that are somehow unchangeable, often seen as biologically given.

What happens if we ask people to list the characteristics of human nature? Usually the result looks something like this:

Greed	Violence
Competition	Intelligence
Ambition	Joy
Jealousy	Aggression

The list usually varies from culture to culture and sometimes from individual to individual within cultures. People listing characteristics of human nature usually make little attempt to make definitive judgments as to what is universal or invariant. Though the list may contain some positive aspects, it is usually loaded on the negative side. On further exploration, many people will explain the negative aspects by noting that these characteristics are typical of most other people, but less so of themselves, while the positive aspects are more likely to be their own characteristics: "I am good" and "They are bad." Social psychologists have labeled this phenomenon the self-serving bias, and it is related to the tendency to attribute other people's (mis)behavior to personal or dispositional factors, while one's own misfortune is assigned to situational, impersonal factors—that is, other people are responsible for their fate, while I am just a victim of extraneous forces. Sometimes, people may admit that the negative dispositional aspects also apply to themselves, but in general individuals describing human nature frequently see themselves as an exception. People will at times explain their own undesirable actions by asserting that "they are only human."

Assigning negative characteristics to human nature rather than to oneself has some practical utility. It provides an excellent excuse for misbehavior. When people excuse their aggressive and competitive, often greedy and dehumanizing, actions, they exempt their own volition and indict "human nature." If it is human nature that is doing "it," then you cannot help yourself. And such acts of human nature are somehow seen as part of our biological makeup, an inevitable, and often deplorable, aspect of being human.

It is only after some probing that the positive side of human nature emerges in everyday discourse. It is often referred to as one's "essential humanity," which refers to the caring, empathic nature of the beast. And it sometimes takes poets and philosophers to extol our humanity, our moral sense, our altruism. In fact, the majority of human interactions are positive or neutral. It may be the very relative rarity of true acts of malice, aggression, and destructiveness that makes them stand out and be memorable, thus distorting our view of the relative probability of positive and negative human acts.[2] In short, in comparison with other animals, "*Homo sapiens* is a remarkably genial species."[3]

Whatever the motivation and basis for the use of "human nature" in the common language, there exists a long tradition that sees human beings as essentially shackled by two kinds of forces: biological and social. Thus, whatever we want to or hope to be, humans are seen as inevitably constrained by their social and biological heritage.

Thomas Sowell, while using the rather cumbersome notion of "visions" to describe world views and ideologies, makes a clear-cut distinction between two opposing views of these constraints: He calls them "constrained" and "unconstrained" visions.[4] The constrained vision is exemplified by Adam Smith and the ascription of self-interest as the basic human motive. The unconstrained vision, for which William Godwin serves as Sowell's prototypical advocate, sees humans as "perfectible" and able to overcome their egocentricity. Visions are nonrational attitudes that exemplify two grand political/philosophical traditions. The constrained vision is of human nature as flawed, seeing humans dependent on social tradition and inherently biologically limited. The unconstrained vision refuses to accepts the limits and sees human beings as inherently perfectible. The former view is conservative (Adam Smith, Hobbes, Hayek), the latter is liberal (Rousseau, Paine, Shaw). Sowell puts Marx in between, seeing the "world constrained . . . , though progressively less so, and eventually becoming unconstrained."[5]

Rather than accept one or the other of these two polar positions, I suggest that a more practical task is to specify those constraints that are indeed inevitable by nature of the kind of earth we live on and of the biological character we have evolved in interaction with the physical and social world. In the process of looking at these constraints, we might be able to discern those constraints that we automatically accept as inevitable but that just might be accepted and propagated because they favor a particular institution, societal structure, privilege, etc. But note that there might be historical constraints that are so long lasting that they have become ingrained in our society and culture and act like biological constraints—and may be as difficult to avoid or change. The task we have set ourselves is to examine in greater detail the nature of the constraints that do in fact limit human possibility, to specify the biological and social forces that limit what human beings can and cannot do or think.

Psychologists have until recently avoided complex questions about the social and biological constraints on being human. The past decade has seen a resurgence of such interests, though old habits still remain. Thus, one book on different models of man (*sic*) presented a variety of philosophical and psychological discussions on the minutiae of psychological theory, but little on constraints or evolution. When these aspects were mentioned, they were typically assigned to the "lower" animals.[6] There are two general questions that one can ask about human nature: What is it—what can it be—in principle? and What are the specific components that make up human beings as we know them today? The first question is the grand one, asked by philosophers and theologians, biologists and psychologists. In the common understanding, not far removed from the same positions phrased in more cumbersome locutions, it usually starts with the phrase: "Human beings are basically. . . ." I try to avoid such grand generalizations and restrict myself to an account of what human beings specifically can do, how they act, and discuss, when available, the reasons for such behavior and actions.

The main dimensions within which we can ask questions about human peculiarities can be briefly described, and in the following chapters I refer to them not in any systematic way, but rather as they arise in specific contexts. The major categories of comparison and questions are as follows:

1. The characteristics and qualities we share with other mammals.
2. The characteristics and qualities that make us different from other animals, such as human language and the pervasive use of consciousness.
3. The evolution of human beings during the few million years since we were differentiated from other primates.

4. The genetic and biological characteristics that are currently distinctively human as a result of that evolution.
5. The social context in which human beings live today, with special attention to the different contexts that determine human nature in different cultures.
6. The social and intellectual history of the species that defines what is thinkable and conceptually possible today, again with an awareness of the fact that such history differs in different cultures and societies, and therefore so does the conception of human nature.

Much of this list consists of constraints on our current conception of human nature. These constraints can be divided into two general categories, which will look familiar.

Biological constraints: The physical, genetic makeup of our species which is, for all practical purposes, the same for all cultures and societies. The biological constraints include such obvious characteristics as our sensory apparatus and its restriction to some (often small) sample of the possible range of visual, auditory, and other sense stimuli that can be perceived. Biological constraints also determine the way aggression and competitiveness can develop or fail to develop. Biological constraints may also include such nonobvious limitations and advantages as object permanence, conceptual range (the character and number of things about which we can think and generalize), and the peculiar development of writing—conditioned in part by the development of the human hand. Much of these arguments here and elsewhere are more speculative than one would like, because, as has been noted, "there are no behavioral fossils." Our genes provide us with the potential to be a variety of different human beings—within the basic constraints. Which of these varieties will be exercised depends largely on social factors.

Social constraints: Those that derive from traditions that we take for granted, social attitudes and habits that determine the structure and content of our thoughts, and historical determinants of what is thinkable about humans and their environment. Our view of the physical universe, our use of Arabic (rather than Roman) numbering systems, our food habits, our specific languages, our beliefs in what is good and evil, and, possibly most important, our view of and actions toward other members of the social system, all of these are a function of social constraints generated by our social context and history.

Finally, a word about the categories of human action and thought that I treat in detail in the following chapters. In contrast to others who have treated human nature within more classical biological categories, I concentrate on those human attributes that seem to set us apart from other animals. Thus I consider consciousness because it apparently plays a more important role for humans than for others animals; emotion because emotional experience seems peculiarly human; intelligence because of the general belief that greater intelligence is what distinguishes us from other animals; sex differences because they seem, in contrast to most of the rest of the animal world, so much more pliable in humans;

values because they seem to be something we have/do that other animals do not; and language and thought because they surely distinguish us from the animal world.

The plan of the book is to start with a general discussion of the various constraints indicated in this chapter. I then present an introduction to contemporary evolutionary facts and theory, together with a review of some attempts to explore the evolutionary or genetic background of human behavior. The next two chapters deal with the complex phenomena of mind—consciousness, emotion, and values—followed by a discussion of the social construction of human behavior, with special attention to aggression. This leads into the problem of human differences, particularly with respect to intelligence and gender. I then attempt to understand human morality and our need for freedom. I conclude the sections on human nature proper with a discussion of human cognitive capacities and language. The book concludes with an Appendix, which illustrates the force of social context; in particular how the thoughts and values that psychologists have about human functions and actions and thoughts are themselves colored by the social fabric in which they are developed.

T W O

Historical and Biological Constraints

∞

> There is a history in all men's lives,
> Figuring the nature of the times deceas'd,
> The which observ'd, a man may prophesy,
> With a near aim, of the main chance of things
> As yet not come to life, which in their seeds
> And weak beginnings lie entreasured.
>   Shakespeare *(King Henry IV, II)*

OUR CONCEPTIONS OF human beings are constrained in a variety of ways—there are a number of unthinkable concepts that are a function of our history, our current social context, and our physical or biological makeup. Constraints have been generally discussed in terms of the latter category—that is, what it is about our biological or genetic constitution that constrains the kinds of things we can do and think. Unfortunately, we are all constrained by the same biological chains, and it is not possible for anyone of us to step out of them and to say: "That is it! That is what humans cannot conceptualize or think about." We have to wait for the proverbial Martians to tell us that. I discuss some of our biological constraints later in this chapter; to begin with I want to stress the social and historical context that determines our conception of humanity.

Myths and Human Thought About Human Beings

Here I briefly outline what I consider to be some of the major guideposts in human history over the past 3,000 years or so. This is in no way a history, nor a review of all the important occurrences during that period. Instead, it is a personal account of a few events that helped shape the kinds of conceptions that humans had about themselves and that still shape our view of humanity today.

I have made reference to unthinkable conceptions of humanity. It is, of course, not possible to determine what it is about our current social context that

9

constrains our conceptions. We can, however, look at human history and the received wisdom and beliefs of a period and consequently make educated guesses about views of humanity that would have been unthinkable in earlier historical times. At any one period in time, the prevailing thought became the conventional wisdom and the easily accepted "truths" about humanity and its environment.

In terms of the early history of humanity, the myths and religions of the times were the primary theories and conceptions of the origin and nature of humanity. That approach prevailed until the beginning of the Renaissance in Western culture—a time when not the first but the most serious attempt to replace theological myths with scientific ones began. In other cultures, such as the Chinese, for example, it is not as clear that religion uniquely determined the common view of human nature, and religious or theological views existed without too much conflict side by side with early scientific ones.

It is obviously not my intention nor my competence to discuss the history of human myths and religions of the world. The development of monotheistic conceptions in the Western world and their myths of the origins of humanity and human character are well known, but it is important to acknowledge the earlier religions as developing the source of common knowledge and the original conceptions of the origins of humanity. Creation myths abound in most religions, and are often redundant and regressive, as in the Greco-Roman, Hindu, and Germanic conceptions which perceived the gods as creating humanity, but being themselves created, for example, by primordial giants. By investing their gods with "human" characteristics, such as lust, greed, jealousy, piety, and love, the identification with the powerful gods was relatively easy, and human characteristics could be seen as godlike. Conversely, the more unappetizing human aspects could be accepted, since they were shared with the powerful gods.

The notion of the single creator, such as the unknowable, unique godhead of the Hebrew tradition and the Old Testament, moves the creator much farther from the common understanding, though he is still endowed with some "human" characteristics such as vengeance and wrath. In that identification with the gods, the older myths also make a strict distinction between humans and other animals. It was not until Descartes (in part) and Darwin that some continuity was generally seen between us and our coevolved planet mates.

Characteristic of Western myths, prior to Greek mythology, as well as of others such as the Egyptian and Mayan conceptions, is the embodiment of the sun and the moon as special sources of power, energy, light, and danger. In part this reflects the realization that there would be no crops without the sun, that its light and warmth were crucial to human existence. In contrast, the moon is often seen as dangerous, the symbol of night when the world is dark and cold and when predators prevail. Altogether, the incomprehensible heavens were seen as determining and directing, and as a consequence astronomy (and astrology) tended to be well developed and to become central to many of the early belief systems.

Crucial to the development of Western thought were the Judeo-Christian traditions that enshrined in the Old Testament a theory of the origin of the earth and people. Relations between humans and animals were spelled out—again in terms of specific differences between us and other animals. But as a theory of the creation of the universe, the Old Testament is not that much different from other creation theories, with all powerful gods creating the earth and its creatures and then peopling it with human beings. All of these stories illustrate the need to know about origins, to explain and understand where humans and their earth came from.

The myths as theories of origins and purpose constrained what was thinkable. It was not possible to have theories of the cosmos like the Einsteinian ones, or theories of natural selection—there was no groundwork laid, nor were any of the relevant observations available or possible. There were, of course, the unusual and adventurous who considered the possibilities of other worlds, other intelligent life, but we are concerned with the common understanding—the range of received knowledge and conventional wisdom.

Of all the earliest quasi-scientific approaches to human knowledge, early Greek thought was probably the most inquisitive, and it started an influential scientific approach. The Milesians (approximately 600–450 B.C.E.) were actively interested in the natural origins of humans. Men such as Thales and Anaximander laid the groundwork for the pre-Socratic tradition in their analysis of the physical world. Anaximander's theory of evolution of humans from fish would later provide Plato with a target in his attacks on prescientific investigations. Others, such as Democritus, analyzed human sensation and the visual apparatus. With Socrates, Plato, and Aristotle we enter the golden age of Greece, and both the pinnacle and the beginning of the end of Greek rationalism and scientific endeavor. It was marked to a large extent by the wide visions of Aristotle (384–322 B.C.E.), his commitment to empiricism, ardent collation of natural knowledge, and investigations in practically all aspects of human, animal, and plant existence. He was preceded by the inquisitiveness of Socrates and by Plato's (427–347 B.C.E.) concern with mathematics, but also by the undermining of the budding scientific tradition by Plato's followers. The concern with positive knowledge in Greece was possibly related to the fact that the Greeks were less sure about their religious myths, less certain of their religious beliefs, and as a result less dogmatic and less proselytizing.

The force of this development in the Greek city-states shaped Western thought to some extent to this day and for many centuries in some detail. For example, Aristotle was to dominate scholarly thought for nearly 2,000 years. The period also produced serious countercurrents to rational scientific inquiry. These trends can be traced back to Plato, his notion of preformed ideas and opposition to empirical investigation, as well as his reactionary views of society, with the notion of established and unchangeable hierarchies. Plato had resurrected a dualism against the naturalistic monism of the Milesians, while also remonstrating against the natural science tradition. In the *Timaeus* Plato decries organic nature as a degeneration of perfect humanity produced by the Creator, and he denounces and degrades natural science.

The Greek tradition of learning and science found another peak before it decayed, and that was when Greek science's central activities shifted to Alexandria, which had become a center of Western civilization soon after its founding by

Alexander in 332 B.C.E. The Alexandrian Golden Age during the second and third centuries B.C.E. was cut short by the disturbances that preceded and accompanied the Roman occupation in the last half-century B.C.E. Whether it was the end of the Golden Age of Alexandria and the eventual destruction of its great library or the influence of antiempiricist views that brought the age of Greek science to an end, it ended by the time Rome had become the inheritor of the mantle of empire and intellectual leadership. During its prime Alexandria had seen such intellectual leaders as Eratosthenes, who deduced that the earth was round and computed its diameter and circumference with astounding precision; Aristarchus, who followed Pythagoras in asserting a heliocentric cosmology; Euclid, who probably wrote his *Elements* in Alexandria; and Herophilus, who founded a long-lasting school of medicine there.[1]

With the destruction of the greatest collection of contemporary human knowledge in the library of Alexandria, the hope for a reasoned approach to an understanding of humanity lay in shambles essentially until the early Renaissance. The loss of that library deprived us of valuable information and, more important, marked the end of the emerging scientific tradition. Alexandria suffered from invading armies from the north and south, and the library (containing more than 500,000 volumes) was partially destroyed in the first two centuries of the current era and entirely destroyed in the sixth century.

I bypass temporarily the events in Galilee, which did not have any major effect for another 200 to 300 years, and which we shall presently encounter. As the Hellenistic period ends and Roman dominance begins, we encounter the solidifying effects of the glory of Rome as it extends the heritage of Greek culture to much of the Western world. Rome did not make any major contributions to human knowledge when compared with its Greek inheritance, but it did provide—for better or for worse—a great unifying force in the Western World and produced great poets and engineers, great and miserable emperors, initiated practical approaches to human existence, such as the memory systems for their orators, and maintained existing knowledge by, for example, stocking the library at Alexandria.

With the fourth and fifth centuries the Christian (Catholic) church becomes a dominant force in Western thought, and we encounter one of my pet villains of human history: St. Augustine of Hippo. At the time, the church fathers—mostly situated in Asia Minor and North Africa—were putting the final touches on church dogma. Up to St. Augustine, the debate was fairly free and open and a pluralistic church was emerging. The episode in Eden was interpreted as an affirmation of human freedom—the ability of man (not woman) to dispute with God, to assert his ability to choose, and even choose to be punished for his transgressions. St. Augustine put an end to this nonsense, succeeded in promulgating the dogma of original sin, and as the first sociobiologist even insisted that the stigma of original sin was transmitted through human semen. Adam's sin (Eve, qua woman, was a minor actor) was ineradicable and transmitted to all humankind. We cannot transcend our basic sinfulness—which makes human nature basically evil and in need of control and renunciation. This is a far step from the freedom advocated by some of St. Augustine's predecessors. The dogma also had its practical consequences for the Roman authorities, since the need for the control of base human instincts made it possible for the Roman state to invoke "biblical" authority for its dominance and power over its citizen.[2]

With the emergence and dominance of church dogma, Western civilization entered the Dark Ages, which, though not quite as dark as generally assumed, still were essentially devoid of any major advances in human knowledge. They were illuminated by contrast in Africa by the scholarship of Arab sages such as Avicenna and by the (albeit church controlled) foundation from the eleventh century on of the universities of Bologna and Paris, followed by Oxford. The highpoint of the scholasticism was St. Thomas Aquinas and his establishment of the still generally recognized philosophy of the church. The general mainstay of knowledge was neo-Aristotelian—the references were Aristotle and the Bible, not empirical observation. The period is well marked by the apocryphal story of the monk who meekly suggested, in the midst of a discussion about the number of teeth that a horse has, that one actually might go out into the stables and *count* them!

With the fifteenth century we enter the next great revolutionary period in human thought—the Renaissance—which changed our conception of human beings in the world and their interactions with one another. Ptolemy's astronomic theory of seeing our earth as the center of the universe gave way to Copernicus. There followed a general acceptance of the heliocentric view—that the sun is the center of our world, and we are but a satellite. Galileo (1564–1642) fathered modern science and introduced its observational basis, and with the acceptance of Copernicus, a fundamental change in the way we construct our knowledge and the way we see ourselves was introduced. We were not—physically—the center of the universe anymore. But there were no changes with respect to the view of humans at the center of the earthly universe.

Other momentous changes affected human knowledge and its dissemination, in particular when Gutenberg invented movable type, though he sold the shop and machinery early on and walked into the mist of history. Among the first major contributors to modern thoughts about human nature was Vives, born in 1492, a predecessor of René Descartes, who speculated about such topics as memory, understanding, and the will. Matters started to heat up in the sixteenth century, with the advent of the revolt against the heavy hand of the church by Martin Luther and the Reformation, but also with the notion of predestination and the eventual Calvinist doctrine of the essential impotence of human endeavor. But sixteenth century scholarship was still to some extent in the thrall of Aristotle: no less than 46 commentaries on Aristotle's *De Anima* alone were published during the 1500s. On the other hand, the century gave us fundamental changes in our perceptions of the world—the change in pictorial representation, in the discovery of perspective as an important new way of seeing and looking.

By the seventeenth century the study of the nature of the human mind hit full stride, particularly with the advent of Descartes (1596–1650). I see Descartes as an ambivalent character on the world stage. His successful enthronement of the immaterial, antimaterialistic soul/mind in Cartesian dualism has confused our understanding of what it means to be human, while establishing a successful cottage industry for philosophers. On the other hand, Descartes made fundamental contributions to science and mathematics, and he reflected the technology of his age when he invoked the clockwork-like working of the animal body. Clockworks were ubiquitous, enjoying great popularity in the seventeenth century, and many exquisite works of mechanical engineering were produced as well. The flavor of the times is reflected by the fact that in the year of the publication of Descartes' meditations (1647), Pascal produced the first working mechanical calculator (as did Leibniz around the same time in Germany).

Descartes was very modern in his desire to model behavior, as seen in this passage: "if there were any machines which had the organs and appearance of a monkey or of some other unreasoning animal, we would have no way of telling that it was not of the same nature as these animals."[3] The passage anticipates modern views, as in Alan Turing's similar comparisons between the digital computer and the human mind.[4] Even though monkeys could be simulated by machines, Descartes asserted that humans would be recognized as such (or not) because the machine: (1) could never use words or other signs for "communicating its thoughts to others" and (2) would not act by understanding or reason. Descartes argued that reason is an all-purpose device, whereas in its absence "organs have to be arranged in a particular way for each particular action." There would not be enough "different devices in a machine to make it behave in all the occurrences of life as our reason makes us behave"[5]

For Descartes, "lower" animals are but mechanical devices, and he argues that one could not tell the difference between real and constructed monkeys. But the difference between humans and other animals is preserved; humans have a soul. Humans are similar to other primates in the construction of their bodies but have in addition the rational mind or soul. They do have free will, rationality, and personal responsibility.

One of consequences of the Cartesian view is that it permits the blurring of distinctions between different kinds of humans. If we postulate human beings who are devoid of the divine soul, we can treat them as we treat animals. Did Descartes make slavery morally possible? Does the Cartesian view justify the distinction among "races"? Does it justify treating slaves, and particularly black slaves, as chattel and animals (as the Greeks and Romans rarely treated their slaves)?

The next two centuries saw major advances in many of the natural sciences following the success of the Newtonian program. With new insights in physics, chemistry, and particularly in geology, we arrive in the nineteenth century and the fruition of evolutionary ideas, culminating in Charles Darwin. Evolution represented the next step from Descartes in our view of the place of humans in the animal kingdom. For Darwin, not only are humans like other animals in body, but we are fundamentally alike, having common ancestors. In preceding centuries we had discovered that the heliocentric view of the world was only part of the story. The sun is center of our system, but not the center of our galaxy, and even the galaxy is not the center of the universe. Most recently, we have encountered an even more radical change—the evolution of the universe itself may be transitory, as it expands from the kernel of the Big Bang, contracts, and then in tens or hundreds of billions of years begins its course all over again. Thus even the "progress" of the universe and of humankind may be repeated over and over again.

What humankind was left with at this point was the notion that we are just another animal among animals on a planet on the edge of the universe, where other beings and intelligences may exist. But we are rational, intelligent animals, determining our own futures, thinking and willing our destinies, and aware of our reasons and motives.

With Marx's economic determinism and Freud's unconscious determinism, even this view came to be undermined. Freud taught us that the rationality is more apparent than real, that we act for reasons or motives of which we may not be, and even cannot be, aware.

And thus we are left with a very different view of human beings and their place in the universe from what we had accepted 3000 years earlier. Now science has replaced religion as the source of myths we accept as plausible explanations of our world. Will something else follow science? What is unthinkable today?

Biological Constraints

It is obvious that most of our physical makeup, our sensory capabilities, and many other human characteristics are the result of evolutionary processes generating adaptations of varying utility to our physical as well as our social needs. These are generally considered to represent the universality in the human species. However, great care must be exercised in determining what is and what is not a universal biological constraint on possible human thought and action. The formula that

$$\text{Species specific} + \text{universal} = \text{part of our genetic base}$$

needs careful consideration.[6] Just because characteristics or behaviors seem specific to humans and are apparently universal does not permit the conclusion that they have a genetic basis. Consider the use of fire and handwriting, both of which apparently fulfill these two conditions but are surely not genetic, though the five-fingered hand—so useful for writing—surely is. Or consider a more controversial topic—the incest taboo. Whereas it is true that the majority of individuals prefer to mate with persons outside their immediate family circle, "majority" is not enough of an argument for a genetic base. More important, however, is the fact that incest is *not*, cross culturally, universally taboo. What is taboo in one society may actually be prescribed in another. It seems to be the case that "[o]ne man's taboo 'sibling' is the same man's wife elsewhere."[7]

If the incest taboo is genetic, why do we need legal prohibitions? It would be like having laws against seeing ultraviolet light. Granted exogamous marriages have genetic advantages, which could be expressed in cultural customs, but it is unlikely that the danger of recessive genes being expressed in incestuous couplings would be obvious to early cultures.

I briefly discuss some of the biological constraints on the development and characteristics of human beings. Several of these categories are taken from the work of Lumsden and Wilson, who have devoted much effort to the discovery of these constraints. However, their claim that it is a "remarkable result" that "mental development appears to be genetically constrained"[8] says either too little or too much. There is no doubt that much of development, particularly that dealing with the physical environment, depends on gene–environmental interactions, but a large part of mental development depends on mainly social factors. Lumsden and Wilson's conceptualization of the important interactions is found in their 1981 book, under the rubric of primary and secondary epigenetic rules. They define epigenesis as "the total process of interaction between genes and the environment during development, with the genes being expressed through epigenetic rules."[9]

The epigenetic rules are "genetically determined procedures that direct the assembly of the mind, including the screening of stimuli by peripheral sensory filters, the internuncial cellular organizing processes, and the deeper processes of directed cognition."[10] These rules embody the constraints that genetic endowment places on human (psychological) development. I now turn to some of these constraints.

Vision and Hearing

What are we able to perceive? Only a limited range of the electromagnetic spectrum is visible, and only a limited range of vibrations is perceived as sound. These limitations are the result of human evolution, but they are not necessarily the best outcome—that is, evolution does not produce the best kind of organisms, just an (often barely) adequate one. But we are the result of prior evolution that has affected the visual and auditory ranges; we cannot choose to change to a broader range. Suffice it to say that in terms of what is "out there," our sensory capabilities are quite limited, and it has been the *social* development of technology and science that has provided us with the tools to extend these sensory capabilities.

Early Spatial Preferences

From birth on, human infants show preferences for certain shapes and for certain scanning strategies, such as boundaries and sharp angles. Such characteristics form the basis for building a repertory of perceptual thought and action. More important is the apparent ability of the young infant to use primitive spatial schemas to develop perceptual categories and to use these schemas as the building blocks for conceptual thought.[11]

Facial Displays

A tradition going back at least to Charles Darwin claims the universality of facial displays (facial expressions) in expressing underlying feelings and emotions. Recent research suggests that whereas there are some such commonalities, the origin and use of facial displays has much to do with the social situations in which they are used and that social motives and early preverbal communicative needs are as important determiners of their occurrence as any possible "emotional" genetic basis. I return to this topic in chapter 6, which discusses emotion.

Early Taste Preferences

From birth on, all infants show a preference for sweet and an avoidance of bitter substances. Other similar taste preferences also appear to be genetically based. These preferences have obvious nutritional and survival advantages.

Phobias

Fears of certain classes of objects seem to be found in most if not all human societies. It has been argued that this is an evolutionary response to ancient threats, such as closed spaces, heights, thunderstorms, running water, spiders, and snakes. However, even the universality of these phobias is in dispute, and one might ask why they have failed to develop in response to other threats such as predators and life-endangering actions of other persons. In addition, some of these phobias are likely to fall under the rubric of generally unusual situations, such as intense stimulation that engenders emotional reactions, which I discuss later.

Face recognition

Humans seem to be better at recognizing faces than at recognizing other regular patterns, and young infants seem to have a preference for the human face when scanning their environment. However, given the frequency with which we encounter human faces from birth on, and that we learn to associate the faces of caregivers with the satisfaction of needs, the genetic basis of face recognition may be questioned. What clearly is the case is that highly integrated perceptual patterns appear to be recognized and reacted to differently than less integrated ones. And this very fact may, of course, be ascribed to the evolved character of the human (mammalian?) perceptual apparatus. Some research has suggested that human face recognition may be an acquired expertise rather than a "genetic" characteristic. Humans apparently become face experts by age 10, and they use representations of faces that differ from the representations used by novices.

Consider the well-known vulnerability of human face recognition to inverted presentation; we have great difficulty recognizing a familiar face if it is shown upside down. It has been demonstrated that the same vulnerability is shown by dog experts in their recognition of inverted photographs of dogs. In other words, the vulnerability is not restricted to human faces, it apparently occurs with other very familiar figures.[12]

Limitation of Conscious, Short-term Memory Capacity

One of the continuing puzzles of human information processing is the limitation of the number of things or items we can (consciously) hold in our minds at any one time. The number is limited to some five events or items, seems to be invulnerable to change, and is surely universal.[13] I shall return to this topic in the chapter on human consciousness.

Mother–Infant Bonding

The fact that children and their mothers establish special perceptual and emotional relationships, though apparently universal but not species specific (i.e., these relationships are also found in other animals), seems to be amenable to an analysis in terms of other psychological variables. The bonding apparently takes place with any consistent caretaker, and the absence of specific bonding seems to be of no particular disadvantage in societies where children are brought up by multiple caretakers and their biological mothers may not even be identified. The importance of the mother in the single mother–child interaction in providing for the needs and wants of the infant, the frequency of exposure to the mother, and the mother's investment in the child as a function of her pregnancy and perception of the child as hers seem sufficient to account for the bonding phenomenon.

Object Constancy

"Object constancy" refers to the observation that adult human beings invariably know that a particular physical object maintains an identity, a permanence in time and space. I know, in the sense that I have an unshakable belief, that the brown shoes I wore day before yesterday and left under my bed are the very same, identical pair that I found under the bed this morning. I know in the same fashion that the car that just disappeared around the next corner has continued its physical existence on the next street and, if I hurry, I would see it being driven down the other street. I also believe that my friend Algernon, last seen a year ago, is the same person (object) now sitting in my living room, despite the fact that he has lost 50 pounds, grown a beard, colored his grey hair brown, and is now wearing spectacles. Objects are permanent, no matter where they may be or how much they may change; they exist continuously in time and space.

Object permanence has been intensely investigated, primarily as a result of Jean Piaget's work. Piaget noted that infants do not display the concept of object permanence until about nine months of age, after the concept of physical causality has been achieved.[14] A number of criticisms have been advanced of Piaget's position, which he based in part on the observation that infants do not— until about 9 months of age—search for hidden objects. One of the most telling and well-controlled experiments that challenges this position has been performed by Renée Baillargeon and her colleagues, who showed that 5-month-old infants consistently reacted differently to a screen moving against an object or apparently moving *through* a hidden object; they found the latter event "surprising."[15]

The argument over the four-month difference is important because it involves the question of what infants need to know—what skills they need in order to display an approximation to the universal adult's transactions with objects. Baillargeon et al. agree with Piaget that a world of permanent objects is not only spatially constituted, but it also displays causal regularities and temporal attributes. In other words, object permanence is "an inseparable aspect of the infant's knowledge of how objects behave in time and space."[16]

The "constraint"—the universal knowledge—that objects are permanent in time and space is a fact that we learn about our universe. It is the behavior of objects in time and space that is incorporated at a very early age in our repertory of knowledge. Absent other information, we would have to argue that in a world where objects do not display such constancies, its human inhabitants would accept the nonpermanence of objects as a universal characteristic. But things may not be that simple. After all, we are dealing with a sentient being that is able to accept knowledge about cause, time, and space. If such Kantian characteristics were not part of our armamentarium, how would we react to permanent objects?

Some of the constraints that are apparently genetic can be classed under the rubric of the "psychic unity of humanity."[17] All human beings learn, as do all other animals. "Learning" includes such a wide variety of different mechanisms that it is obvious to psychologists and biologists alike that "it" is not a proper locus of evolutionary study. That normal human beings all learn complex thoughts and actions and learn them well (however different they may be) suggests that learning should probably be assigned a heritability of 0—that is, there is no effective variation within the species. In other words, different human groups do not show different variations in "learning." One subsection of learning is the capacity for inference-making, which seems to be present in all members of the species. As a result, similar cultures, faced with similar problems, will arrive at similar solutions.[18]

How do the categories that we have just surveyed help us explain human behavior in mature cultures? Except for the incest taboo, the ubiquity of learning, object constancy, and possibly the facial expression of "emotion," very little is explained. Most of these categories have little effect on adult social thought and behavior. Which of these categories, or some of the additional ones listed by Lumsden and Wilson, such as color vocabularies or "acoustic behavior," permit any significant predictions about human social organization, our representation of our social world, or how we order and organize our world? I doubt anybody would claim that they significantly affect human problem solving or social interactions. Surely, the perceptual variations in the visual arts and cuisines, admittedly within the confines of the "constraints," are nearly as great within cultures as they appear to be between cultures. And phobias are rare enough to play too important a role in social organization or in problem solving. Thus, I will have little occasion to refer to these constraints in what follows; I will, however, introduce some new ones and emphasize factors that are more directly involved in our social existence.

Other Important Constraints

Among the characteristics of humanity that I have neglected in my foregoing list, several deserve a more extensive treatment, which they will be given in subsequent chapters. They include the following:

1. The pervasive perceptual ability to respond to the fact that the situation one is in and the world one lives in is actually the same as it usu-

ally is or is, in fact, different. This "difference/sameness detector" is important to such diverse human characteristics as stranger anxiety in children, the origins of emotions, and even the source of human values.

2. The role of consciousness as a semiautonomous system that aids in monitoring and selecting from the multitude of events that happen both external and internal to our bodies.

3. The role and development of human language and the question of whether it is the product of a unique evolutionary event or the grand achievement of a variety of different cognitive skills.

4. The ability of human beings of nearly identical physical makeup to adapt to environments of dramatically different characteristics, varying from the heat and dryness of the desert to the cold and wet of the Arctic. We adapt to these environments rather than avoid them, which could have been a possible evolutionary choice—that is, to stay huddled together in the cradle of humankind, Africa.

And finally, an example that I do not discuss further: the importance of the specific kind of gravity that we encounter on our planet. We have developed characteristics of which we are rarely aware that have conformed to the requirements of that particular gravity. These adaptations are an excellent example of environmental–evolutionary interactions. If earth's gravity had been less or greater, we would be very different animals—we would have different sizes and limbs, and would have developed different tools, different modes of travel, and different forms of work.

Evolution and the Genetics of Behavior

⚭

Evolution . . . is a change from an indefinite, incoherent homogeneity, to a definite, coherent heterogeneity.

Herbert Spencer

THE GENETIC CHARACTERISTICS of humans cannot be considered without referring to the evolutionary events that have given rise to them. Evolution is sometimes seen as somewhat mysterious, sometimes as irreligious, sometimes as competing with a vague "creationism." I will not spend time discussing the latter, unacceptable fantasies, but rather will present a short review of the basic postulates and facts of evolutionary theory.[1]

General Principles of Evolution

Evolution functions like any well-established scientific theory—that is, as a combination of observations (facts) and explanatory statements that is (temporarily) satisfactory in providing us with an understanding of a range of phenomena. We start with two questions that elicit the range of evolutionary theory: What is the evidence for the evolution of animals? What are the observations to be accounted for?

1. Life has a continuous history of change through time. That change is characterized in part by adaptation to new or changing conditions and in part by flaws in animal design. For example, the limbs of reptiles are rather clumsy and obviously not the "best" solution to the problem of locomotion. The reason for this flaw is that the reptiles' limbs evolved from fishy fins. But in the process of change, new conditions demand solutions (e.g., locomotion on solid terrain) us-

ing the materials at hand. Nature has to work with whatever is available. Thus, one of the substantiating sets of facts are the imperfections produced by evolution. It is a blind process not necessarily ending up in perfect organisms or the best adaptation.

2. The geographic distribution of animals must also be considered. The earth has not remained the same over its history, and its inhabitants reflect the progression of geological changes. The evolving geological landscape has important general influences on the evolution of animals; as geology changes, so does the ecological niche for any particular animal and, as a result, the adaptive, selective pressures on its evolution. Geography (or ecology) exerts pressures on evolution that generate similar solutions to similar problems. For example, there are three different classes of mammals: placentals, marsupials, and monotremes. All have a common ancestor in some Ur-mammal and share the common characteristic that they nurse their young with milk and also have hair. But placental young mature inside their mother's body, marsupials mature inside a pouch, and monotremes lay eggs. Faced with similar ecological niches, however, we find both marsupial (Australian) and placental (e.g., European) wolves, mice, anteaters, and squirrels. But note that marsupials are more closely related to each other than any marsupial is to its analogous placental animal.

3. Vestigial organs and functions mark the history of earlier forms of species. These are signs of evolutionary history, and often practical imperfections. They include the Moro reflex in infants, the appendix, and the nonfunctional toe in horses. They are vestiges of ancestral functions and organs.

These changes, geographical observations, and evolutionary imperfections were all challenges to the kind of theory that Darwin proposed. How does that theory proceed to help us understand the plant and animal world and its sometimes peculiar characteristics?

Darwinian Theory

The best account we have that can encompass all these and other observations is the theory of natural selection proposed by Charles Darwin. Others had proposed bits and pieces of evolutionary theory, but Darwin put the grand design together, and except for some changes brought about by subsequent research, it still remains the best way to account for the change over time in the composition and character of the animal world. The major points of the contemporary version of the theory are:

1. Genetic characteristics of a species change from generation to generation. These changes may be attributed to (a) mutation due to naturally occurring radiation and other little understood factors and (b) recombination of genes.

2. Natural selection assumes that (a) there are differences among individuals within a species, (b) there are variations among members of species in their inherent or genetic makeup that produce some individuals that are better able to adjust to the demands of their environment and their needs, (c) the number of offspring is larger than the number that survive, (d) on the average, the survivors

possess those characteristics (adaptations) that enable them to live a more productive life in the sense of being more effective procreators, and (e) selection pressures at any one time and in any one environment will differ for specific characteristics—that is, will be more important for some characteristics than for others.

Given these genetic changes and their effects on some of the offspring, we
arrive at changing individuals as a function of their interaction with each other
and the environment. Yet it should be pointed out that Darwin was unaware of
Mendel's theory of gene transmission and believed that inherited characteristics
were determined by the blending of inherited traits. According to that view,
offspring represent a blended combination of their parents' characteristics. This
is an illustration of the original theory of evolution being wrong in one aspect,
even though the facts seemed to fit. The erroneous aspect is then replaced by
another, in this case Mendelian genetics, and the theory as a whole is improved. Another erroneous attempt was the proposal by Lamarck, who tried to
understand evolution by asserting the generational transmission of acquired
characteristics.

Evolutionary theory sees evolution as purposeless. Evolution develops out of
the struggle of individuals to survive, and the survivors pass their genes on to
their offspring. There is no goal involved; evolution does not guarantee a better
and better animal or an improvement of the conditions of life. Thus, there is no
real distinction between "higher" and "lower" animals—all are specific examples
of some degree of adaptation to their history and their ecological niche. For example, the cockroach is essentially the same animal it was hundreds of thousands
of years ago because its characteristics were adaptive then and are now—there is
always enough food for cockroaches. On the other hand, the woolly mammoth
did not survive as the environment changed.

The Selfish Gene

One of the challenges to evolutionary theory was the observation that people
can perform altruistic acts. Why does an animal do something that is good—
something life-preserving and protective—for another animal? When an animal
produces a warning cry at the approach of a predator, it protects another member of the group so it can flee, but the crier exposes itself to increased danger.
Where is the individual struggle for survival? It is generally accepted today that
Hamilton's account of inclusive fitness best explains these observations.[2]
Hamilton's argument is based on the fact that, on average, two siblings will have
50% genetic similarity, and similar genetic sharing occurs for other relatives.
Therefore an act of altruism that ensures the survival (fitness) of a relative will
ensure the survival of a large part of one's own genetic material. Thus, birds that
produce a warning cry that exposes them to a predator also protect their own
genes in their kin. Warning is an apparently unselfish act that in fact promotes
the spread of one's own genes. The genes that "survive" include those that allowed the animal to produce warning cries. However, we must distinguish this
kind of altruism from much of human altruism, which is determined (encour-

aged and defined) by cultural and social factors. Altruistic saints are moved by their beliefs, faiths, and commitments, not usually by a likelihood of protecting their genetic material.

In connection with this genetically determined unselfish behavior, some explanation is in order for the widely used term "selfish gene."[3] In everyday parlance the word "selfish" implies some intention or action initiated by agents concerned primarily with their own welfare and good. Genes do not, of course, have any intentions or any concern with their welfare. We describe genetic action as selfish because genes survive if the behavior and characteristics they control is such that it is more likely that the gene will be passed on to the next generation. Thus, a gene is "selfish" because its totally automatic effects are such that they ensure the survival of the gene, including the "altruistic" gene and its "offspring." The misuse of the psychological concept of "selfishness" within evolutionary theory has led to many misunderstandings about selfishness being a trait of human individuals as a result of a "selfish" gene. No such generalization is warranted or defensible.[4]

The intentional implication of the selfish gene becomes particularly inappropriate when we consider that the competition for individual advantage also includes cooperative behavior. If some genetic characteristic makes for cooperative behavior and mutual aid such that the individual carriers of the "cooperative" gene derive some reproductive benefit, then that cooperative gene is selfish in the same sense as one that produces an actual competitive struggle. Survival and reproductive advantage can come from cooperation as well as competition.[5] In considering the immediately observable adaptations of an organism, it is useful to remember the distinction between the *genotype* of an organism, that is, the sum total of its genetic makeup, and its *phenotype,* the "visible" aspect of the organism that is the result of the interaction of the genotype and environment and which is, after all, the part of the organism that interacts with the world.

Adaptation

Does natural selection produce ideal organisms? And if not, why? There are a number of forces working against perfection:

1. Natural selection works on the entire organism, not just parts. It is the ability of the organism to survive, not the sometimes temporary advantage of a particular characteristic that determines survival and selection. Each part's function in the adaptability of the whole organism needs to be considered. Thus, some animals will survive because their total makeup is better suited to their niche, even though some brothers or sisters may have a characteristic that provides a temporary advantage in solving a particular problem.

2. Natural selection is not a process for perfection; it is a design process that works on what it is presented with, however imperfect. For example, elks' antlers are selected primarily for sexual display and some competitive struggles, but they have also made the animal slower and more vulnerable to predators.

3. Environments, both geographic/meteorological and "social" (in terms of neighbors and predators), change frequently, and often more quickly than evo-

lutionary processes. As a result, the design of a particular animal may be quickly outdated. Only if environments—in every sense—stayed constant over millions of years could one arrive at an optimal design for an organism.

4. Gene transmission often occurs in "clumps." Neighboring genes are often involved in transitions and mutations. Thus, some genes may be selected but are not selected *for* a particular adaptive reason.

Adaptation has generally been used in two main senses. On the one hand, it is used to explain both cultural and biological change—to insist in the extreme case that all change occurs because it better fits the organism to handle the stresses and strains and the goals and rewards of its environment. On the other hand, it is used as a purely biological concept—to explain current genetic makeup in terms of its value in producing reproductive success. There are a number of different arguments that have been made about the uses and misuses of the concept of adaptation, and I shall concentrate on those issues central to the arguments of this book.

To use adaptation in its extreme sense—to say that whatever the current state of an organism, it is best adapted to its current niche—is circular. It leads to the reasoning exhibited by Voltaire's character Pangloss that this is "the best of all possible worlds." The circularity asserts that a characteristic or trait exists because it is adaptive, and it must be adaptive because it has been "selected." What is needed, whenever possible, is a test of adaptiveness and selection, independent of its current existence. One must also resist the temptation to use intuition in place of empirical evidence. It is not acceptable to be "content . . . with measures that we intuit as being related to fitness."[6]

There are many cultural, as well as biological, features of human beings that are in fact maladaptive, certainly in terms of reproductive success. For example, the modern use of contraceptives does not contribute to adaptation or to reproductive success. Human culture may also be adaptive for some members of the society, but not for others. The prevalence of class-structured societies provides excellent examples of cases where a cultural innovation is well suited for the well-being of one class, but is detrimental to another—and even destructive to the reproductive success of the disadvantaged class. It should also be remembered that just because a particular phenotypic expression is actually or potentially adaptive does not imply that it will be the target of evolutionary selection. Selection typically operates on the activity of the organism in its niche and less frequently on isolated individual traits. Selection operating on species as a whole is a contentious issue.[7] Can evolution work toward the survival of the group or the species, or is the survival of the group a function of the survival of the individual who benefits from the particular characteristic and thus contributes to the group? We do not need to decide on this specialized issue, except to say that most selection probably occurs at the individual level. Finally, a trait may be currently perceived as maladaptive (e.g., that some human emotions, such as anger, are unnecessary leftovers of our prehistory), but that does not mean that such perceptions are necessarily correct. Manifest (cultural) judgments need to be analyzed and tested before we can easily dismiss some human trait as unneeded or even destructive to human happiness.

At the same time, just because something is hardwired or genetic does *not* mean that it is necessarily good and beautiful and "adaptive." This adaptationist view has been described as Panglossian. It ignores the facts that much that is genetically determined has emerged from functions unrelated to its current use, that some of it is fortuitous, and that yet other functions have long lost their worldly relevance. Unfortunately, this is not "the best of all possible worlds," and evolutionary processes cannot be relied on to produce the best available human machine. Adaptation and selection are the major selective processes, but they are not foolproof. In addition, evolutionary adaptation does not (usually) at a single leap generate a single intricate machinery, such as a "language acquisition device." Most changes tend to be small and cumulative and diverse. As a result, even "hardwired" modules are likely to be the result of many different evolutionary pressures and adaptations, and different modules are the product of different processes at different times. So in the absence of behavioral fossils (except for aging academics), many claims can be made for the evolution of behavior without fear of contradiction.

Preadaptation or Exaptation

The argument that evolution is indirectly relevant to everything we do but not directly relevant to many things we do only restates the notion that much action and thought arises out of a complex background and is not, by itself, the result of natural selection or adaptation. There is a more general and important illustration in the history of evolution—the phenomenon called *preadaptation*. Briefly, preadaptation deals with those cases where some particular trait or structure, arising out of the usual selection processes, may become useful for some other function of the species later in its history. The term has the unfortunate implication that the characteristic in question is somehow preordained for its subsequent role, and Gould and Vrba have suggested the alternative term *exaptation*, which carries no such implication.[8]

My argument about the complexity of many human traits and characteristics is directed, in part, at the likelihood of exaptation being at work. In chapter 6, I argue that exaptation is at work in the development of human emotion, and I extend that argument to the case of the human desire for personal and political freedom in chapter 10. In general, exaptation tells us that human traits arise as complex products of our evolutionary history, with evolutionary development taking advantage of existing characteristics that are not, initially, relevant to the observed ones.

In general, preadaptation arguments address issues that have bedeviled current as well as early evolutionary theory. How do adaptive traits originate, how do complex structures arise in the first place—so that natural selection can do its work and select those individuals who show the trait?[9] Or, to use nineteenth-century language, how do we account for the incipient stages of useful structures? Specifically, of what possible use is one-tenth of an eye or one-half of a wing, and if they are of no use, how does the structure develop to its full utility if its "incipient" stages could not possibly be the locus of natural selection? An early solution—that the new structures arise suddenly and in full panoply—not only was unlikely, but also anti-Darwinian because it bypassed entirely a basic Darwinian mechanism, that of a slow accretion of useful structures by natural selection.[10]

Darwin's solution, presented in the final edition of the *Origin of Species,* was the principle of functional change in structural continuity. The principle asserts that the function of a structure may change as it evolves; that is, it is not necessary for one-tenth of an eye to be an organ of sight, or for one-half of a wing to be an organ of (impossible) flight. Rather, the function changes as the structure evolves. This principle was called preadaptation because it refers to the fact that there is some adaptation prior to the adaptation finally displayed.

I describe below some ingenious solutions to the problem of the evolution of the wing. But first we need to realize how precarious a concept like exaptation is. If evolution itself often tempts both expert and layperson to concoct "as if" stories, then this is even more tempting in the case of exaptation. The tendency to develop a reasonable-sounding story of previous functional uses of structural precursors is obviously strong, and, conversely, evidence for exaptive function is much more difficult to collect or document than is evidence for "straightforward" evolutionary processes. I am particularly aware of this problem because I argue in the pages which follow that many of the most interesting and complex "human" characteristics are not the result of direct selection of the structure or trait in question but instead are more likely to be the complex product of preadaptive or exaptive processes.

The case of wings is a good example of hard work and ingenious science producing a more than reasonable story of exaptation. The case of animal flight has probably attracted more attention than any other candidate for preadaptive explanations, partly because of the apparent inutility of partial wings and partly because the unlikely direct evolution of wings has been one of the main targets of antiscientific attacks on Darwinian theory. Early attempts at explanation included a variety of candidates for early function, among them their use as sexual attractants or stabilizers of some sort. The "stabilizer" explanation was used extensively in the case of insects, and it is for insect wings that one candidate for early function—thermoregulation—was developed in a scientifically persuasive fashion. The investigators were Kingsolver and Koehl, who performed a series of ingenious experiments as well as applications of mathematical models to look at both the thermoregulatory and the aerodynamic properties of wings. What they found was that, as the wing size increases, it has increasing thermal effects (e.g., cooling of the body), but that at a certain wing size there is no additional thermal effect. What is crucial, however, is that they demonstrated that below a particular wing size (for a given body) there were no significant aerodynamic effects and above that size there were increasing aerodynamic effects. What Kingsolver and Koehl demonstrated is the actual point in the possible evolution of wing size where the shift from an early function (thermoregulation) to a later (and final) function (aerodynamic efficiency) might take place.[11]

Another theory of the evolution of insect wings has been proposed by Marden and Kramer. These investigators were observing the surface-skimming behavior of stoneflies and suggested that the initial short, stubby appendages (evolved from gills) were useful for skimming along the surface of water. These rudimentary wings could evolve into larger wings that would at first support gliding and eventually flying. Again, a series of ingenious experiments and theoretical work on the importance of muscle power and wing effectiveness produced important support for the theory.[12] It is this kind of elegant experimentation and modeling that is most likely to develop persuasive arguments for particular functional shifts in the process of exaptation and final adaptation. I am not able to present such arguments here for some of the cases that interest me, but I shall try to be cautious in making generalizations and urge the reader to do the same.

The evidence for preadaptation and the apparent slow cumulative development of the parts of the brain that support language argue for a relative slow, often happenstance, development of modern human beings. Even bone and muscle development seems to have gone along with changes in social organization (from scavenging to planned hunting). It is therefore highly likely that most characteristics of modern humans did not appear full blown, as I shall argue later for the development of language. As it was recently argued elegantly, "the modern package didn't arise full blown. It came in pieces."[13]

The Evolution of Humans

Any discussion about the evolution of human behavior must be constrained by a consideration of human evolution in general. The path of evolution that is specifically human starts with the branching off of the hominid line from the ancestors we share with the other primates. Much exciting work and many discoveries have clarified and sometimes confused the specifics of that evolution.[14] For our purposes, one can assume that the break occurred more than 5 million years ago and that *Homo habilis,* our most humanlike ancestor, lived some 2 million years ago in Africa, followed by the new *Homo erectus,* who spread out from Africa into Europe and Asia. No earlier than a half million years ago, there appeared the first member of a group of hominids called generally early *Homo sapiens.* Eventually *H. sapiens* proper emerged in Africa no earlier than 200,000 to 250,000 years ago. In all that time, behavioral evolution had adequate opportunities to develop unique human characteristics. I illustrate one approach to our understanding of that evolution with the physical evidence for speech and language.

The Physical Evidence on Speech and Language

Evidence for the physical equipment necessary for speech and language comes from various areas of investigation. Much evidence has been collected from the fossilized remains of early human craniums. Using endocranial casts, Phillip V. Tobias has suggested that evidence for linguistic ability, possibly true spoken language, but more important, the potential for symbolic communication, first appeared in *H. habilis* about 1.8 million years ago.[15] Previous to this, endocasts of *Australopithecus africanus* and *A. robustus,* the precursors of *H. habilis,* show development on the left hemisphere of the brain in an area corresponding to Broca's area in the frontal lobe. Broca's area is important for the fine motor control needed in speech and it acts together with a region called the supplementary motor area (or SMA), located within the left hemisphere. Because of the position of the SMA, it cannot leave a trace on an endocranial cast. Damage to Broca's area in modern humans results in inability to coordinate motor aspects of language, but leaves comprehension intact. With the appearance of *H. habilis,* brain size increased significantly, and endocasts show a bulge corresponding to Wernicke's area in the left superior temporal lobe. This area is involved in speech comprehension and the symbolic aspect of language. Thus far, the endocranial evidence has suggested that Broca's area was developed prior to *H. habilis,* while Wernicke's areas, associated with the symbolic aspect of speech and necessary for comprehension, first appeared some 1 to 2 million years ago with *H. habilis.* Compared to other estimates, endocranial evidence has postulated a long period of development for human language; currently, most estimates have suggested that human speech and language appeared about 100,000 years ago. But Tobias's evidence suggests at least that humans had the capacity for the *precursors* of speech long before language itself developed. What we can tell from the physical evidence is that a potential capacity for language was present, but this does not tell us whether *H. habilis* was practicing verbal language or its precursor, sign language.[16] When and how the physical equipment for producing spoken speech first appeared have also been the subject of intense investigation, with current estimates that the full laryngeal equipment was not in place until less than 200,000 years ago.[17]

The Last Steps

True *Homo sapiens* emerged some 200,000 years ago, and brain size has remained fairly constant since then. The emergence of language itself is usually placed at about the same time, as is the anatomical change that permitted the pharyngeal structure necessary for human speech. The major changes in human society, which included sophisticated artifacts and clothing, complex dwellings, the beginnings of art and ritual, can be placed at 40,000 to 50,000 years ago. Given that language did not emerge miraculously (see chapter 11), it seems defensible to argue that full human language developed 100,000 or more years ago and became fully operational as a cultural transmitter about 50,000 years ago in the Early Upper Paleolithic period.[18] It is interesting to speculate that 40,000 to 50,000 years ago our ancestors might have been endowed with the same cerebral and physical capacity as that which supports twentieth-century literature, physics, and philosophy.[19]

The limitations of natural selection apply also to human characteristics, including that most intricate of evolutionary creations, the human brain. Our brain is the result of millions of evolutionary events over a period of millions of years, but it is not—no matter how highly we think of ourselves—an ideal device or design. It has, in fact, built-in limitations. It is the best we can get at the present time, but there are constraints that prevent it from being better than it is. First, the human head at birth cannot exceed the maximum diameter of the birth canal, and human general structure and some requirement of locomotion limit the size of the female pelvis, and second, the brain uses already a very large part of cardiac output.

Early representatives of premodern humans were the Cro-Magnon and the Neanderthal, subspecies of *H. erectus* and the precursors of *H. sapiens*. Either Neanderthals lost out in competition and died out and were replaced by modern humans (*H. sapiens sapiens*), or the two subspecies cross-mated and modern humans are a mix of the two.[20] In any case, Neanderthals disappeared as a separate subspecies some 30,000 to 45,000 years ago. These developments took place after *H. erectus* moved out of Africa, and most of the evidence for the Neanderthal and Cro-Magnon conflict has been found in Western Europe. The dominant view today is that modern humans (or their direct ancestors) developed in southern Africa. A minority view holds that precursors of modern humans spread very early from Africa and then went through parallel evolutions. At present, we must wait for more evidence before one of these views can be chosen as correct. However, the force of argument is against the likelihood of separate and identical (!) developments and for the notion of an African development of *H. sapiens* and their subsequent dispersal.[21] The most recent summary based on converging evidence suggests that current dispersed human lineages split off from a common human ancestor in Africa probably no later that 110,000 years ago and no earlier than 200,000 years ago.

The so-called modern races probably did not develop until 100,000 to 200,000 years ago and were adaptations to the local environment, with some characteristics selected for particular local weather and geographical conditions. Thus, skin color is a response to hot and cold environments. A similar selection may have occurred for eye color, but the latter may be linked or clumped genetically with skin color and may not by itself be the result of natural selection.

Some 15,000 to 50,000 years ago (depending whose side one takes in a current scientific debate) humans arrived in Siberia and moved across the land bridge to North America and then spread out in North and South America. Recent research has shown that by about 3,000 years ago, culturally and socially rich societies had been established in Central America. This was followed by the significant and impressive cultural developments of this migration in South America about 2,000 to 3,000 years ago, with the great South American societies established 1,500 to 2,000 years ago.

The difference between primitive humans mastering fire or, some time later, drawing bisons on the walls of caves, and those who master calculus at age 15 or compose symphonies, is primarily a difference due to cultural, social, and historical—not biological—evolution. The immense difference between human social organization 50,000 ago and that which we enjoy and suffer today will not be understood by appeals to our biology.[22]

Universality and the Argument for Genetic Similarity

The argument about universality usually hinges on the observation that all human groups or all hominids or all primates display a certain characteristic or behavior. It is then assumed that this (often not necessarily correct) observation permits the conclusion that the characteristic observed has the same function for all the individuals or groups in question, or, to go a step farther, that the characteristic has the same evolutionary history in all the cases, or, in the extreme, that it somehow expresses the same underlying genetic makeup. An illustrative counterexample[23] should suffice to undermine most of these inductive leaps. Sharks, porpoises, and the extinct ichthyosaurs all have streamlined bodies and similar fins. But consider the fact that sharks are somewhat primitive chordates, porpoises are mammals, and ichthyosaurs were land-living reptilians. Studying the evolution of these three animals as a group makes no sense; certainly, the evolution of their fins has no common biological history. To be sure, fins are an excellent solution to the problem of mobility in a watery environment, but that speaks at best only to the problem of function. If we want to study the evolution of fins, we will have three quite different stories to tell.

In general, two species that display identical (or highly similar) physical characteristics or behaviors may do so because one of the two descended from the other, because both of them descended from some common ancestor, or because both are facing similar problems of "engineering," of solving certain tasks imposed by their environment, often under very different conditions. The problem is described in evolutionary theory as the distinction between *homology* and *analogy*. Homologous development arises out of common ancestry, common genetic constitutions, while analogous solutions are independently arrived at in response to common environmental constraints. The same argument applies, of course, to cultural evolution. Just because different societies seem to have similar (or even identical) rituals, beliefs, or modes of living does not—in any sense—suggest that these similarities have developed with similar histories or for similar reasons. Monogamy, incest taboos, and handwriting, among other human endeavors, may have developed as reasonable engineering solutions for certain problems under quite different conditions, just like the fins. If we wish to understand their origins, if we wish to speculate about their biological or genetic basis, we need, if at all possible, to look at their history and development under each of the conditions (societies) in which they appear. If we want to assert common human biological causes for universal traits, we must make sure that they are in fact universal, that it is unlikely that they have appeared for different social or biological conditions in each of their instantiations, that we have some reasonable story to tell of how their evolution might have taken place, and that no reasonable alternative solution to their appearance has equal claims on evidence and explanatory power.[24]

The Role of Change

Even within biological arguments, we need to be cautious about assigning all change, all evolution, to natural selection. In addition to the driving force of natural selection, such phenomena as genetic drift and the incidental effects of selecting traits other than the one under consideration need to be seriously considered.

In general, in contrasting cultural and biological change, it is useful to pay attention to three major characteristics that differentiate the two: rate, modifiability, and diffusibility.[25] The *rate* of change due to cultural evolution is extremely rapid when compared with biological evolution. We are, in the twentieth century, pretty much the same animals that Hamurrabi, Homer, Moses, and Assyrian peasants, Greek soldiers, or Hebrew fishermen were. We may not be that much different from inhabitants of the caves at Altamira. However, we talk, count, construct, travel, pray, and dance quite differently; our skin color and hair color are differently distributed across the earth than they were then, and we know more about ourselves than those biological contemporaries did.

Both simple and complex characteristics of culture may be radically *modified* in relatively short time periods. The monotheism of a significant portion of the Hebrews changed 2 millennia ago and did so within a period of less than 100 years. The dominant governing class changed, as a result of social revolution, in such diverse countries as France, England, India, Russia, China, and Kenya. Similarly, views about the intelligence and abilities of groups of people changed radically in decades. Dominant American views of inferior African-American intelligence in the mid-twentieth century find their parallels in identical views around the turn of the century, but applied to Irish or Jewish intelligence.

Cultural traits can be diffused by a variety of methods, including teaching, coercion, and imitation. Thus, the evolution of cultural traits is complex, existing in a network of communication groups and subgroups that exhibit both positive and negative feedback among groups and societies.

Sociobiology and Behavior Genetics

At the simplest level, sociobiology is concerned with genetic or evolutionary codeterminants of human social behavior. Nobody can argue with the aims of a discipline circumscribed in those terms, and knowing about the relative effects of nature and nurture on our behavior and institutions is of great importance and interest. However, the term sociobiology has become a label for a point of view that stresses these biological characteristics at the expense of social, historical factors. The arguments are typically based on an extension of knowledge about other animals and animal groups. I discuss some aspects of sociobiological speculations that are related to my major concerns here; in a later section I analyze the bases of behavior genetics, often based on twin research. The major criticisms of sociobiology have been directed against the incomparability of animal and human behavior, the lack of evidence in the human realm, the facile use of analogy, and the lack of mechanisms that demonstrate how one might go from the gene to complex and often symbolic behavior.[26]

The sociobiological position may be addressed from the ideological point of view, but it can also be examined critically from a primarily factual position. For example, Washburn notes in connection with Konrad Lorenz's work on aggression that "bits of behaviors from oddly selected animals" are used "to suggest causes for important human social problems."[27] The most serious problems are that frequently animal behavior is used to "explain" human behavior. Washburn concludes that "[s]erious comparisons require careful, detailed analysis and experiments wherever possible."[28] And similarly, Frank Beach referred to the futility of comparing selected aspects of human and animal behavior "when the sole basis for selection is a common label referring to superficial similarities."[29] Like Washburn, he stressed that the important "first step is separate analysis of the behavior *in each species independently.*"[30]

These criticisms are not directed against obscure aspects of the sociobiological view, but rather against some of its major tenets. For example, George Barlow advocates the use of animal studies because they "allow us to abstract principles, by analogy, as to how the behavioral system of an organism is adaptive."[31] While using vague analogies, one may apparently at the same time reject anthropological evidence that negates the desired general principles of human behavior. Barlow deplores the fact that "[f]or all but the most general conclusions, across animal species and human, you can always find some obscure human society that negates the generality."[32] The implication here—and in other rejections of such evidence—is that an "obscure human society" can be dismissed as a distributional quirk that does not invalidate the generalization to be defended. The basic argument is that just as a six-fingered human hand is a random biological event (which it surely is), so is, for example, a nonaggressive or matriarchal society. The difference, of course, is that the biological quirk appears once and then in often unpredictable ways, while the "obscure" society involves hundreds or thousands of individuals and, barring *cultural* interventions, usually survives for centuries. It would surely be frivolous to argue against a genetic behavioral trait by pointing to one or two individuals who fail to show it, and the single compassionate, cooperative Samaritan in a competitive, aggressive society may indeed be an accidental product of that society, but the lasting cooperative and collective efforts of several hundred people suggest that, whatever its origins, societal structure may not reflect some hypothetical genetic inheritance.

One of the problems that recurs in trying to sort out the gene–culture problem is the identification of the behavioral trait—the unit of behavior—that is actually genetically determined or guided. I return to this issue in a discussion of specific issues, especially of aggression—that is, what behavior is it in "aggression" that arises out of our genetic makeup? But I want to raise here a more general point. To start with, one can surely agree with the sociobiological axiom that we may be "confident that evolution has relevance to behavior—all behavior."[33] Surely, humans could not—in the simplest case—act at all if they did not have the physical equipment with which to act. In that sense morphology is prior, and the major characteristics of our physical equipment are surely genetically determined. But to say that evolution has relevance to all behavior does not imply that all behavior is (in any direct sense) genetically determined. All humans use their hands to wash themselves, primitive counting is almost exclusively done with fingers, the easy manipulation of feet is used in dances in all societies, but nobody would argue that evolution has selected hands or feet for washing, counting, or dancing. In these cases, physical traits that were selected for some particular function have become useful for some other, often quite unrelated, ones.

What quickly becomes apparent is that the analysis of complex behavior is often quite vague, and the assignment of cultural or genetic evolution is at best imprecise and usually impossible. The more complex the behavior, the more difficult it becomes to talk about evolutionary relevance. What genetic factors are "relevant" to a preference for rock music over a Bartok quartet, or the choice of bitter over sweet chocolate, or holiday preferences for mountains over the seashore? Once behavior becomes complex, the genetic determinants tend to be overwhelmed by the cultural ones, and the link between genetic determination and the observed behavior becomes weaker. The approach that I take in this book is to concentrate on the simplest acts and behaviors that may reasonably be said to be specific evolutionary products and to proceed from these to explore what other more complex behaviors may be influenced by them. Technically, this is a concern with proximal mechanisms that may be responsible for distal characteristics.

A serious sociobiology needs to be concerned with proximate mechanisms that are responsible for the phenotypically observed complex social behavior. Philip Kitcher has argued effectively for such an approach, and Gould has stressed the importance of both proximate causes (investigated by developmental biology) and functions (the "final causes" explored by evolutionary biologists) in understanding any adaptive structure.[34] Kitcher has noted that the social behaviors of interest to sociobiologists are best viewed in terms of their proximate, underlying causes. Underlying proximate mechanisms are likely responsible for a variety of behaviors, and it is unlikely that natural selection has acted directly on any (or all) of the individual tokens of such behavioral varieties.

Behavior Genetics and Adoptive Twin Research

In recent years, the more extravagant speculations of early sociobiologists have given place to dedicated research on the genetics of human characteristics. The empirical analysis of possible genetic roots of human behavior is of central importance and an important antidote to the excesses of some sociobiologists. In what follows, I concentrate on some shortcomings of adoptive twin research, not to denigrate the work of behavioral geneticists, but rather to offer friendly criticism. One approach has been to study similarities in the behavior of identical (monozygotic) twins, examining the degree to which twins show similar or even identical behavior in their early and later lives. Another approach is to examine the behavior of identical twins separated at birth and adopted and therefore reared in different environments. The claim is that these studies hold genetics constant while varying the environment. I believe it is important to examine these claims in detail because of the importance of the questions asked and the claims made by this kind of research.

First, let me illustrate the general problem of dividing observed behavior into components of genetics and environment. All sides to the various arguments about nature and nurture agree that genetic endowment and environmental, situational influences generate a product; they interact, are interdependent, and rarely if ever are behavior and action determined by social or genetic factors alone. The goal of sociobiologists and behavior geneticists alike is to divide observations into two neatly defined categories: heredity and environment. Before I address the vexing problem of defining environments, let me illustrate the notion of the nature–nurture interaction. I use a somewhat different, though relevant, use of the concept of "product." Imagine that I want you to guess by what multiplicative process I arrived at the number 24. Unless I give you some additional hints, there is no way you can arrive at a correct solution. You just don't know whether it is 2×12, 3×8, and so forth, unless you have further information. And similarly, all we are given when we observe behavior is the sum or product; there is no easy way to find out the contribution of either our environment or our genes to our behavior—all we have is the product. What to do?

Nature has provided us with an interesting natural experiment that might well solve part of our problem: the existence of identical twins. For the purpose of this discussion we can assume that identical twins are, in fact, totally identical in genetic makeup and environment up to the moment of their delivery.[35] What might we be able to do with nature's gift to research? The argument goes that if we place the twins in different environments and if their behavior then turns out to be different, we can assign their differences to nurturing, environmental differences; if they stay identical or at least similar in action and behavior, then we could assign the source of their action to their genetic makeup. I will not go into the details about how such a partitioning can be done and some of the statistical techniques involved because I argue that under present circumstances such experiments are fundamentally inconclusive. In order to illustrate that it may be impossible to differentiate between "what" (heredity) and "where" (environment) under current constraints, consider the following thought experiment in perfume marketing.

I have started a new perfume firm with the appropriate hoopla, endorsements, and packaging, and have decided to start with 20 different scents, ranging in their appeal from some called "Passion," "Lust," and "Desire" to others with names such as "Peace" and "Tranquility." I want to know whether these scents will sell as well in different markets or whether I should target different markets with different scents. So I prepare identical pairs of shipments of all the 20 scents and one batch goes to stores in London, the other goes to Paris. Within weeks the results are in: the perfumes sold equally well in both places. So, I conclude triumphantly at a board meeting, it isn't *where* you sell, it's *what* you sell that makes a difference. One smart-aleck young member of my board who has taken experimental design courses in college insists that I cannot say that without providing a better difference between the two markets—they are too much alike. Alright, I say, let's try Amsterdam and Beijing the next time. We do, and this time, the results are a little more equivocal—some of the scents do better in one place, some in the other, and after an expensive statistical analysis by a group of consultants, we are told that both *where* and *what* make a difference, maybe. However, we do not know why the Dutch like some scents and the Chinese like others. And now my young board member comes up with another question: How do we intend to use this information—just in Amsterdam, Beijing, London, and Paris? No, I say—and I am getting irritated—this will determine our worldwide strategy. But, he says, how can you generalize to the whole world when you haven't sampled some of the more obscure places in the world, and you still don't know why certain perfumes do well or badly in some places. In fact, you ought to place your identical batches for sale randomly in places where perfumes are sold, then you might have some idea whether *what* and *where* are different sources of market appeal. After that you will still have to determine *which* perfumes to market preferentially *where*.

The parallel with twin research is obvious; the problems are the same. When identical twins are raised in different environments—specifically, when the twins are adopted and placed with different families—what ranges or varieties of the environment are being sampled? Typically few, because twins tend to be placed in similar environments.[36] The lore about adoption agencies is that they try to place twins in similar environments (and until recently some adoptive placements had even been restricted by law) or that twins are placed in the first available families. In either case they are placed in "similar" families belonging to the same culture, and consider how unlikely it is that one of the twins will be placed with a family living under the poverty level and the other with the family of the president of a major company. But even if adoptions were random within a particular town, state, or country, the similarity between environments would still be there; two households in the United States are more similar than one in the United States and the other in Siberia, North China, Corsica, or Papua New Guinea.[37]

Similar problems of environmental sampling (and norms) apply to a different paradigm of twin studies. In addition to examining similarities between identical (monozygotic) twin pairs, a control group of nonidentical (dizygotic) twins is included. Both types of twins are unseparated; that is, they are reared in the "same" environment. The argument is that in one group genetic endowment is held constant and environment is varied, while in the other group both endowment and environment are allowed to vary. If the behavior of the identical twins is more alike than that of the nonidentical twins, then one concludes that genetics is more important than environment. While in the previous paradigm it is argued that the environment (in the two adoptive homes) is different, in this case it is argued that both identical and fraternal twins share the same environment. However, this assumes that identical twins, who look alike, are treated the same way as fraternal twins, who do not.[38] In fact, it is likely, and there is some evidence, that physical similarity will influence the degree of similar or differential treatment.[39]

Not much hope can be offered if we consider the most sophisticated possible twin study—examining adopted identical and adopted fraternal twins. If the environment of adopted twins is as similar as I suggest, then there is little room for environmental effects to show themselves in either group; second, the nonidentical twins are not, after all, random selections of genetic endowment; and third, we do not know whether adoption agencies handle the placement of identical twins the same way they do the adoption of nonidentical twins.

To conclude from any of these studies that *what* (genetics) is more important than *where* (environment) is surely extravagant. But the situation gets even worse if we want to make general statements about nature and nurture rather than (as the example to this point should suggest) restricting them to a single country or culture. If one wants to make statements about nature and nurture in general, about such important topics as aggression or intelligence, one needs to place the adoptive twins in homes randomly selected all over the inhabited world. To make the results even more generalizable, one ought to have twins from all possible genetic compositions. The latter is in principle impossible and the former is in fact impossible. But short of coming up with some solution to these problems, the results of twin adoption studies need to be considered as initial hypotheses rather than as hard data.

To illustrate these points briefly, I offer some examples.[40] A study of television viewing not only illustrates the problem of twin studies, but also the need for the kind of proximal analyses that I advocated in the previous section. Plomin et al.[41] computed the percentage of variation in television viewing due to genetic factors. They concluded that 45% of that variation was genetic. However, that estimate is arguably unreliable, but whatever that estimate is, more important is the fact the "television viewing" is very much a "distal" phenomenon. As Prescott et al. note: "Does viewing time reflect passivity, need for stimulation, or some other characteristic of adaptive significance? If 'passivity' is the behavior under biometric scrutiny, then why not measure it in several reliable ways? Can we make the standard assumption that TV viewing is a homogeneous psychological characteristic that is well understood?"[42] In contrast, it is surprising how little concern has gone into Plomin et al.'s original discussion; they entertain no mediating hypotheses and are content to note that "no obvious physiological mechanism suggests itself as the intermediary of genetic influence."[43]

One study of identical and nonidentical twins has (not unexpectedly) attracted the attention of the sensationalist press, since it concluded that there was a "strong influence of genetic factors in the etiology of divorce."[44] The finding was that the risk for divorce is greater for individuals whose identical co-twins have been divorced than it is for individuals whose nonidentical co-twins have been divorced. The data were obtained from an unknown percentage of the available twin cohorts. We do not know what proportion of the twin pairs actually responded to the inquiry and, more important, the method is subject to obvious sampling biases: for example, what is the likelihood that somebody would reply to such an inquiry given that there is or is not a history of divorce in their family? A number of possible factors that are never entertained by the authors can be adduced for their findings. For example, identical twins look more alike than nonidentical ones and are therefore more similar in attractiveness. In our society physical attractiveness is associated with the availability of alternative partners, and such availability is associated with divorce; therefore, divorce risk among identical co-twins should be greater than the divorce risk among nonidentical twins.[45] In addition, even if correct, the conclusion about a genetic basis for divorce risk is severely limited by the fact that divorce and attractiveness interact differently in different cultures (e.g., when marriages are arranged). Also, one cannot speak of a genetic risk of divorce as if there were no interaction between heredity and environment, and given the use of a single (Western) culture, such an interaction could not emerge.

Is there any possible solution to the problem of environmental sampling? What is needed is a more careful analysis of environments; we have no taxonomy of environments at the present, and an understanding of the complexities of gene–environment interactions depends on being equally precise about both factors. The interaction of environment and genetic endowment is at best difficult to disentangle; it becomes even more difficult when one tries to define and specify the functional environment. For example, a recent study on the heritability of schizophrenia started with the fact that monozygotic twins are likely to be concordant for schizophrenia at about 50% (compared with dizygotic twins at less than 20%). Davis and his co-workers[46] then further explored the environment of these monozygotic twins, in this case the prenatal environment. Twins may prenatally share one placenta (monochorionic) or may each develop in a separate placenta (dichorionic). The latter have a different environment, particularly with respect to shared fetal blood circulation. It turned out that the concordance for schizophrenia for monochorionic pairs was 60%, whereas for dichorionic pairs it was 11%. Thus, with the identical genetic makeup of these twin pairs, the prenatal environment was crucial in determining the eventual development of schizophrenia.

If the prenatal environment interacts with genetic disposition, consider how much the personal and historical environment interacts with social dispositions. What will be needed is a multidimensional approach to an understanding of "environments." First, we need to know which environmental factors affect specific behavior and how they do so (that is particularly important with respect to questions of intelligence to be discussed in a later chapter). We need to consider the summing and interactive influences of such things as parental education and economic status, the individuals' education and their social support. Second, we need to know how various environments are distributed in the human habitat (the world at large). The choice of environments may be limited if one wished to make generalizations only about nature–nurture interactions in New York or in France or in Western industrial countries or in preliterate societies, and so on. Third, we also need to determine how best to combine various environmental measures into a single useful measure of environmental efficacy. Once such groundwork research has been completed, one could compare the difference in the environments of adopted twins with the actual variability available in the target population to which one wishes to generalize.[47]

Finally, I return to the need for understanding more about the genetic component being investigated. Are the twins in adoptive homes a random sample of possible genetic human constitutions? Obviously not, but they may give us some indication which restricted genetic sample is being studied. In searching for the genetic contribution to the nature–nurture interaction, we need more considered searches for the proximate, underlying characteristic responsible for the behavior in question. If one were always to take the simple road of trying to find genetic bases for complex social behavior, one would inevitably "discover" such eventually ludicrous sociobiological phenomena as the size of one's bank account turning out to have one of the highest genetic heritability indices.[48]

The most interesting aspect of human evolutionary history may not be any single social trait, but rather the fact that we are extremely social animals. With few exceptions we live in social groups, depend on social cooperation for our food and shelter, and, because of the great vulnerability of our young, devise social means for taking care of and protecting our offspring to adulthood. I noted in chapter 1 the predominance of positive interactions among humans, our misperceptions of human malice notwithstanding. It is the evolution of this human geniality that may be a more interesting puzzle than the putative genetic basis of divorce or TV viewing.

Having concluded some preliminary discussions of the framework, social and biological, needed for some understanding of human characteristics, I now move on to specific topics, starting with an approach to the notion of "mind."

Minds, Bodies, and Schemas

∽

Instruct them how the mind of man becomes
A thousand times more beautiful than the earth
On which he dwells . . .

William Wordsworth

PROBABLY THE MOST elusive of all characteristics that we assign to our nature, to what it means to be human, is the human mind. It exists in all kinds of shapes and disguises, names and allusions, whether it is the English *mind* or *spirit,* or the German, Spanish, Italian, or French version in such diverse forms as *Geist, mente, ésprit,* and *Psyche.* The mind is possibly our grandest invention, the intangible core of humanity. And it has found its way into the languages of both philosophy and psychology—for philosophers an often confusing label for consciousness; for psychologists a bone of contention for decades of study. Psychologists have long understood that it is futile to try to use the common language as a vehicle for scientific abstractions and generalizations or to explicate psychology by trying to define what common conceptions of emotion or intelligence might be. Given that sophistication, it is surprising that many psychologists (and certainly the philosophically inclined ones) are still trying to determine what mind is. In the common understanding, "mind" has a variety of referents, as shown in such expressions as "I can't keep my mind on it," "I am going out of my mind," "She has a fine mind," or "It is a matter of mind over body." Just this small sample seems to refer to such different events or concepts as attentional capacity, rationality, intelligence, and the influence of some noncorporeal entity. The same kind of confusion exists *pari passu* with the term "mental." It is not surprising then that "mind" has achieved such an exalted status and that folk models of mind have drawn attention, inasmuch as they determine to some extent the way in which people consider their own and others' thought processes and construct their view of the world. In any case, "mind" and "mental" are not directly observable; they are inferred from observations. And they are used freely to refer both to the self and to others. So much for the philosophers' arguments concerning the existence of minds other than one's own. They are all inferred, and the remaining question is merely whether the inference is based on the observation of the self or of others.

The sense in which many philosophers use the term "mind" is highly restrictive, but it could provide some consensus in that it uses the conscious contents as the referent for mind, including such events as conscious thought, perception, and feelings. Mind, then, more or less equals consciousness. If we do not wish to restrict "mind" to conscious contents and processes, the remaining general sense appears to refer to an agency that is responsible for (determines) human thought and action. We can be all inclusive and accommodate most common language usage by using "mind" to refer to the inferred mechanisms and processes assigned to the active human being—processes that construct the contents of consciousness, determine our feelings, attitudes, and beliefs, and guide our intelligent (and not so intelligent) actions.

A position that sees mind and consciousness as independent, though related, concepts is implicitly present in many psychological discussions of mind and has been at times explicitly defined. For example, some 45 years ago, Karl Deutsch suggested that "*Mind* might be provisionally defined as any self-sustaining physical process which includes the seven operations of abstracting, communicating, storing, subdividing, recalling, recombining and reapplying items of information."[1] Such an approach to mind as a collection of mechanisms is also implied by some philosophers, even though they are preoccupied with the mental functions of consciousness. For example, one philosopher notes that "most of the mental phenomena in [a] person's existence are not present to consciousness" and "most of our mental life at any given point is unconscious."[2] And finally, if mind is the repository of perceptual, cognitive, behavioral mechanisms, then it can also be argued to be the function that is performed by the brain. If "mind is what the brain does," then similar relations can be seen in the form and function of other human organs. A related sentiment is echoed in the assertion that "the mystery of the mental is no more a mystery than the heart, the kidney, the carburetor or the pocket calculator."[3]

It should be agreed that the most important mental processes take place unconsciously. This has been a contentious issue in the past, but the advent of contemporary cognitive psychology and the cognitive sciences should put that argument to rest. The notion that much of human thought and action is motivated, constructed, and directed by processes that are not available to conscious knowledge or inspection has been with us in its strongest form since Sigmund Freud and, in experimental psychology, since the Würzburg school. The former demonstrated the weakness of rational thought in understanding human behavior, the latter showed that the steps necessary for even simple problem solving are not accessible to consciousness. The best example of the power and ubiquity of unconscious processes and knowledge is found in language. We do not learn how to use grammar in primary school. Average 3-year-olds speak and understand their native language without difficulty; they know no grammar—consciously. The structures (rules) that generate their grammatical behavior must be present somewhere because they construct their grammatical speech. And since these rules are not consciously available, we call them unconscious—no more, no less. The same is true of much of our daily problem solving, memorial, and decision-making activities. Anyone who still claims that the concept of the unconscious is unnecessary must provide an alternative explanation of the phenomena observed.

Minds and Bodies

What is the relationship of such a "mind" to the physical substrate in which it operates? To get at this puzzle, which has been called the mind-body problem, we can look at the relationship between physical data and mental data. How can we envision the brain as the generator of mental functions? There are specific, and sometimes very precise, concepts associated with the function of larger units such as organs, organisms, and machines, concepts that cannot without loss of meaning be reduced to the constituent processes of the larger units. The speed of a car, the conserving function of the liver, and the notion of a noun phrase are not reducible to internal-combustion engines, liver cells, or neurons. *Emergence* is a label that has often been applied to the new properties that arise from large assemblies. But rather than saying that the new properties emerge, it might be more parsimonious to insist that different entities have different functions. The mind has functions that are different from those of the central nervous system, just as societies function in ways that cannot be reduced to the function of individual minds.

Much of the difficulty that has been generated by the mind–body distinction stems from the fact that mind and body are discussed in terms of ordinary-language definitions of one or the other. Because these descriptions are far from being well-developed or well-understood systems, it is doubtful whether the problems of mind and body as exemplified in everyday language or as developed by the philosophers are capable of providing significant insights. If we restrict our discussion of the mind–body problem to the often vague and frequently contradictory speculations of ordinary language, then, as centuries of philosophical literature have shown, the morass is unavoidable and bottomless.

For example, we could, in the ordinary-language sense, ask how it is that physical systems can have "feelings." Usually the question is phrased as if "feelings" were a basic characteristic of the mental system instead of one of its products. The report of a feeling is a complex outcome of a complicated and complex mental system. Not only is the experience of a feeling a product, but its expression, through a language system, is the result of mental structures that intervene between its occurrence in consciousness and its expression in language.[4] Thus, any question about the relationship of feelings to physical systems is premature. The question about the relationship (the correspondences) between physical systems and feelings requires that we know what the physical system and the mental system are about, and that the theories and functions of these systems are unequivocal in their prediction and specified in great detail as to their structure. Until these goals are achieved, scientific questions about the mind–body problem in general are premature and irrelevant. What we can do at present is ask fruitful questions about specific mental events and their underlying physiological implementation. That has happened in some areas of neuropsychology and neurophysiology. But these partial insights into correspondences cannot and are not intended to solve the mind–body problem as it has been handed down from the Greeks.

In short, there is no special mind–body problem. It is one of many examples in which the interface between systems of explanation, theory, and conceptualization requires bridging concepts, acceptance of discontinuities, or admissions of ignorance. With the development of the explanatory systems that abut such an interface, the nature of the problem changes to the extent that, in some cases, reduction (explanation) of one system in terms of the other may become partially or even wholly possible. The mind–body problem has been the subject of special attention because it is of immediate phenomenal relevance to an active, cogitating, theory-building organism—the human being.

Current attempts to relate the mind–body problem to computational solutions harks back to the seventeenth century, when Descartes implicitly supplied the answer when he claimed that, for lower animals, machines could be designed that would be indistinguishable from the "real thing." His test for the adequacy of such machines in the age of clockworks was the forerunner of a similar test that the mathematician Turing developed for the age of computers.[5] Recall that Turing's test is similar to Descartes's implied answer for lower animals: if we can design machines that think (aloud) and act like humans with ideas (minds, souls) and bodies, then we will have moved closer to solving the ancient conundrum. When a machine becomes indistinguishable from a human actor, it will represent a reasonable approximation of how humans work.

In summary, the useful notion that the mind is what the brain does refers to the fact that we usually assign observed or implied or subjectively reported behaviors of humans to some intervening "mental" set of variables. At least since the end of the nineteenth century, and surely for most of the twentieth, the term *mental* has been applied to conscious and unconscious events and to a variety of theoretical machineries, ranging from hydraulic to network to schematic to computational models. In recent years, computational models have achieved a unique status in the history of science as a variety of philosophers and cognitive scientists have acted as if the millennium had arrived and the final model that intervenes between the brain and behavior has been found in the computer analogy.[6] I now turn to the question about the kinds of contents that inhabit the mind that direct, construct, and represent our behavior, thought, and action.

The Nature of Schemas

The notion of the schema has been with us at least since Immanuel Kant, who gave us a prototype that is still valid. In his description of the schema of a dog, for example, Kant described a mental pattern "that can delineate the figure of a four-footed animal in a general manner, without any limitation to any single determinate [i.e., concrete] figure [that] experience, or any image that I can represent . . . , actually presents."[7] Thus schemas are the basic building blocks of everyday thought; they are the major representations (and abstractions) of both our cultural and our individual experiences. We have schemas of people, animals, landscapes, of how to cook, how to write, and so forth.[8] A schema is a coherent—that is, unitary—representation that organizes experience. Schemas are not carbon copies of experience, but abstract representations of experiential regularities. Schemas range from the very concrete, involving the most primitive categorization of perceptual experience, to the very abstract, representing general levels of meaning such as "love" or "justice." Abstract schemas subsume more concrete schemas; the resulting structure is hierarchical. Schemas are built up in the course of experience and interaction with the social and physical environment. Schemas are built up slowly as we experience the world around us, and as we age our schemas become both richer and more numerous. Thus, our first experience with an automobile may be just a schema of a Ford Fiesta, but with additional experience we build up a generic schema of cars, and eventually such a schema will tell us what to expect when, for example, somebody tells us to meet a car outside our house. But in addition to a generic schema of cars, we also develop generic schemas of particular types of cars (such as Fiestas) and specific instantiations of particular cars (such as our own, for example). Schemas organize and interpret our world, and they organize experience in that current encounters are defined and interpreted in terms of the schemas laid down by past similar and cognate experiences. Currently, active schemas define what we are likely to see, hear, and remember and also determine what we are unlikely to hear or see. So we note the "time" when looking at a clock in a public square, but are unlikely to see (process) the precise form of the numerals. New information activates and constructs relevant schemas that in turn organize our experience of the world.

The activation of schemas proceeds automatically from the most concrete to the most abstract relevant schemas. At the same time, and also automatically, activated high-level schemas pass activation to lower schemas (top-down processing), which constrain further perception. Expectations are those elements of schemas activated by top-down processing that are not directly supported by input evidence. Expectations influence what will be attended to by influencing the ease with which new evidence may be interpreted. Expectations are not met when evidence is not found for all or part of an activated schema; the result is incongruity in schematic processing. This incongruity is one of the causes of autonomic (sympathetic) nervous system arousal, in part to prepare the organism to cope with a changing environment. For example, when I walk into a shoe store, I expect to see (find evidence for) shoes; I do not expect to find shoes displayed in an ice cream shop, but I am likely to be "aroused" if I find no shoes in a shoe store and shoes rather than ice cream in an ice cream shop.[9]

Schemas are not rigidly bounded representations but are best seen as dispositional. That means that we do not have hundreds and thousands of schemas stored in our mental system, but rather we have a disposition to construct a schema when the situation demands. What happens then is that currently available information constructs from observed and stored features (stored as a result of previous experiences and widely distributed in the mind/brain) a particular representation that responds both to the immediate information and to the regularities (schemas) generated by past events. Because schemas develop through experience, it is important to consider how they change. The single most important description of such a development has been provided by Jean Piaget, who analyzed the process of schema development in terms of *assimilation*, the integration of new elements or input into existing schemas, and *accommodation*, the change in schemas caused by input that cannot be assimilated. Assimilation refers to one end of a continuum where new information and new events are assimilated into existing schemas without generating any important changes in the prototypical relevant schema. When the new information cannot be assimilated, we refer to the other end of the continuum—accommodation—which implies a change in a schema in order to integrate information that could not be assimilated by current variables.

I stress the importance of schemas because they are the best available system of representation of past experiences, but also because they provide a simple way of showing how our past, our memories, influence our current actions. We usually interpret current conditions, situations, and events around us in terms of our past experiences with and attitudes toward similar events and situations. Furthermore, the main stock of our schemas that interpret the world around us are created, directed, and colored by society, by the culture in which we live and have grown up. One of the illustrations how such cultural experiences and expectations influence our thoughts and images is shown in the Appendix, where I show how our contemporary culture colors the kind of psychology we are likely to construct.

Consciousness

⚯

A spider conducts operations that resemble those of a weaver, and a bee puts to shame many an architect in the construction of her cells. But what distinguishes the worst architect from the best of bees is this, that the architect constructs his structure in imagination before he erects it in reality. At the end of every labour process, we get a result that already existed in the imagination of the labourer at its commencement. He not only effects a change of form in the material on which he works, but he also realizes a purpose of his own . . .

Karl Marx

PROBABLY THE MOST obvious, yet puzzling and mysterious, characteristics of human existence is our consciousness.[1] We are conscious of the world, of our thoughts, of others; we are not just convinced but certain that other people are conscious in ways similar to our own experience. And yet while practically certain of other consciousnesses, we seem to have no way of showing that they exist, nor are we at all sure whether other animals can be or are conscious. In this chapter I explore some of these puzzles.

Philosophers and psychologists have been fascinated with questions about human consciousness throughout the ages. The interest waxed and waned at times, but always arrived at similar questions: How does consciousness work? How does it function? How can one get something that ephemeral from a material body? In contemporary psychology, speculations about the role and functions of consciousness arrived only recently. Most of the excursions into consciousness in the immediately preceding years had been primarily in the traditional philosophical mode—that is, atheoretical—without any real concern about how consciousness fits into the general mental apparatus.[2] On the other hand, and relevant to the present computer-oriented climate, consciousness is not yet "computable," but that does not mean that its understanding is not crucial for an understanding of the human mind. What follows is a continuation of my own efforts to place consciousness, what George Miller called "the constitutive problem of psychology,"[3] in the context of its function in human thought. It is an attempt to define the

role of consciousness in human thought and to confront the skills and characteristics that it confers on human beings.

I use *consciousness* in terms of immediate experience, whether it is the experience of current extra- and intrapsychic events or is reflective. The contents of
consciousness are indeterminate; they are not scientifically observable. Verbal descriptions of conscious states are verbal because that is the only way we can (here
and now) talk about them. On the other hand, conscious states are subjectively
obvious and certainly not exclusively verbal. Another important aspect of conscious experience is that observations or reports of conscious states or contents
change that content. The main reason for this effect is that the capacity of consciousness is limited and the consciousness of our self-observation "displaces"
other contents. I discuss this limitation later; for the present it should be kept in
mind that we are usually conscious of a single (though rapidly changing) scene,
that commentaries on our experiences replace those experiences, that associated
thoughts occur in series and not simultaneously, that—in brief—we cannot think
(be conscious) of more than one thing at a time. I start with a distinction among
four general positions that have been taken toward conscious phenomena:
epiphenomenal, directive, intentional, and constructivist.

Consciousness as epiphenomenon. This position assumes that some mental
processes and contents are consciously accessible but that their conscious status
is not generally relevant to ongoing mental processes and, in particular, does not
interact with such processes. Nothing different would happen if there were no
conscious contents and processes. Psychologists and cognitive scientists of varying persuasions have been found this position attractive. An excellent illustration
is the position taken by Thagard, who suggests that "[c]onsciousness may be like
the heat or the hum or the smell of the computer."[4] In other words, consciousness is a side effect that is of no particular importance in understanding human
thought and its functioning. But then Thagard hedges the epiphenomenal stance
by suggesting that consciousness may have some functions for resolving some
conflicts, for planning, or for pedagogical or social purposes—a rather surprising juxtaposition that seems to destroy his original argument. In their preoccupation with the computer analogy, many cognitive scientists have become uneasy with consciousness as an aspect of mental life. In part because the problem
of computer consciousness is so daunting, and at the very least complex (though
not difficult for science-fiction writers), some philosophers and others have become closet epiphenomenalists, refusing to assign to consciousness any function
in mental life.[5]

Consciousness as directive or executive. The assumption here is that the conscious
status of some mental contents has a special importance, usually by directing or
organizing mental contents. Conscious contents, according to this view, make
possible the manipulation of other mental contents. Consciousness is the "operating system" of the brain; it is a "computer control system" that directs short-
term and plans long-term activities when the system is awake.[6] These and other
proposals imply a rather complex homunculus—a little person inside the mind
that does the operating, deciding, directing, and planning. But the question arises
of who or what programs these systems?

Consciousness and intentionality. The third general position on conscious-ness sees subjective experience as a necessary condition for intentionality.[7] Intentionality may be briefly and inadequately defined as the "aboutness" of mental states—the notion that mental states have intrinsic meaning (semantic contents). The relation of intentionality to consciousness arises out of the asser-tion that the ability to have subjective, conscious experiences is necessary for "understanding," for imbuing mental acts with meaning. One should, however, consider that in some cases the subjective experience accompanying intentional acts may be no more than a conscious gloss on the action and not a necessary condition for it. The identification of intentionality with consciousness is car-ried to an extreme by those philosophers who identify mind with the contents of consciousness.

Consciousness as a functional and constructed device. The constructivist position states that conscious contents are constructed from unconscious ones; they may be identical with them but usually are not. Marcel has introduced the notion that conscious contents make sense of as much of the available evidence as possible (but within capacity limits).[8] Not unlike the executive position, the construc-tivist view sees consciousness as a serial device interacting with the parallel ar-chitecture of the mental system. It derives the functions of constructed con-sciousness from psychological rather than from computational considerations and lessens the "homunculus" functions of consciousness. The conscious repre-sentation has more or less automatic effects that interact with decision processes, memory functions, and so on. This position sees consciousness as functional in that conscious processes fulfill some mental functions without which the organ-ism could not operate in its present form and for which it would have to find other ways of implementing their function.

A Constructivist Approach to Consciousness

Consciousness is constructive—that is, the contents of consciousness are *con-structed* out of available (activated) unconscious contents; they are rarely identi-cal with specific unconscious structures.[9] Such a view asserts that most of our mental armamentarium is represented in unconscious structures and deposito-ries of perceptions and memories, and that access to these unconscious mental contents is parallel, occurring in several underlying structures all at the same time. Ongoing consciousness respond to momentary needs, demands, and tasks; they are not the result of unconscious material being pushed, elevated, or illuminated in consciousness. Consciousness is not a threshold phenomenon, with uncon-scious contents becoming conscious when a threshold of excitation, strength, what have you, has been exceeded. In other words, there is not an identity be-tween what was unconscious and is now conscious, but rather the conscious ma-terial is constructed out of available unconscious elements.[10] Current experi-ence and needs, among other things, activate a variety of different unconscious materials, and phenomenal experience is a specific construction of activated schemas in response to the requirements of the moment. Specifically, schemas are made active whenever some event in the environment or some memory or thought process makes contact with the schematic representation. Thus, seeing a lion activates lion, zoo, and cat schemas (among others), and these are then available for conscious constructions. We are not conscious of the process of activation or of the constituents of activated schemas, but the schemas that are used in constructive consciousness must be adequately activated and must not be inhibited. One of the consequences of this position is that the same concatenation of unconscious mental representations may give rise to different conscious contents as a function of specific situational requirements.

The Functions of Consciousness

Talking about the functions of consciousness brings up the several uses of "functionalism," the most prominent of which are the following: Functionalism$_1$—the use most frequently adopted by philosophers and cognitive scientists—argues that the study of the "functions" of the organism (and its mind, etc.) can be carried out without reference to the specific underlying neurophysiological hardware, typically by the manipulations of symbol systems and by complex computation and by stressing the input–output functions of the system. It is how the brain/mind functions that is of concern, not what its functions are. Functionalism$_2$ is the simplest sense and is used in sensory psychophysics when particular mathematical functions are used to describe variations in experience as a function of variations of the sensory stimulus. Functionalism$_3$ is best represented by the "functionalism" of the Chicago School of psychology around the turn of the century (Carr, Dewey et al.). It was concerned with observable behavior, its effects and its evolutionary "functions." Its secondary emphasis of describing behavior "as a function of" some external events (e.g., McGeoch, and also the psychophysics of functionalism$_2$) can be seen as a forerunner of American behaviorism. Functionalism$_4$ has been primarily used by linguists who wished to contrast their concern with interactive semantic, syntactic, pragmatic, social, psychological, etc. functions of language with formalist and modular views (e.g., Chomsky, Fodor). My own approach comes closest to functionalism$_4$, stressing various functions of consciousness, i.e., asking what consciousness is *for*, but without neglecting its physical (physiological) representation when appropriate.

We are customarily conscious of the important aspects of the environs, but never conscious of all the evidence that enters the sensory gateways or of all our potential knowledge of the event. This implies that any current conscious content is momentarily "important," but it does not, of course, imply that processing or reacting to organismically important events is always conscious. Conscious constructions represent the most general interpretation that is appropriate to the *current* needs and scene in keeping with both the intentions of the individual and the demands of the environment. In the absence of any specific requirements (internally or externally generated), the current construction will be the most general (or abstract) available. As needs and requirements become more specific, conscious contents become more concrete. For example, when trying to remember a name or phrase, we may start with general specifications that often become very specific, eventually descending to the level of phonology and spelling. For example, I try to remember a former student's name: It was short. Was it Ted? Not quite; more like Fred. Fredric? No! Ah, it was Frederick!

The higher-order structures that often mold current consciousness tend to be rather abstract schemas. The instantiation, the "filling out," of these structures may take on a variety of different forms, technically a large number of different variables and values of variables. Conversely, such structures maintain few default values. In other words, they are not active unless they take on specific evidence, specific information from the intra- or extrapsychic environment. In watching a play, one of the structures that constructs consciousness requires some information about plot features (who does what to whom?) and some identification of the dramatic personae. But there is no default value for the plot at this level—I do not usually assume that it is a murder mystery or a Shakespearean historical play if not enough information is available.

Consciousness participates in learning, and while not all new learning is conscious, the acquisition or restructuring of knowledge and most complex skills requires conscious participation. In the adult, thoughts and actions are typically conscious before they become well integrated and subsequently automatic. Learning to drive a car is a conscious process, whereas the skilled driver acts automatically and unconsciously. It follows that, apart from responding to momentary needs, conscious representations more often reflect those mental and behavioral events that are in the process of being acquired or learned and less often reflect the execution of automatic sequences.

The sequence in learning from conscious to unconscious is not ubiquitous. The sequence may be reversed in the infant[11] and apparently is reversed in simple adult functions, such as in perceptual learning and in the acquisition of simple skills (e.g., balancing, whipping cream). I do not mean to restrict automatic learning to simple skills; it is just rare that complex skills are acquired entirely without conscious participation by the adult. The reason is that many of the components of complex skills are available in automatic packages that are then concatenated into a more complex sequence. Beginning automobile drivers "know" how to step on pedals, turn a wheel, flip a switch, and they are usually instructed how to use these skills in the more complex sequence of driving a car. Skills learned unconsciously may subsequently be represented in consciousness. In addition, shifts from unconscious to conscious processing occur frequently. For example, the pianist will acquire skills in playing chords and trills and in reading music, which are at first consciously represented, but then become unconscious. However, the analytic (conscious) mode is used when the accomplished artist practices a particular piece for a concert, when conscious access becomes necessary to achieve the proper emphases, phrasings, and tempos—that is, changes in the automatic skills. A current conscious state will be changed if it does not account for (make sense of) the available evidence. When the environment is constant, we respond to internal demands for conscious constructions. Daydreamers are unaware of their surroundings, until such time as a shout or a raindrop demands to be accounted for in the stream of consciousness. A similar scenario occurs during conscious troubleshooting, when some automatic sequence of action or thought fails and we immediately become conscious of the current scene (e.g., when driving a car and a brake fails or an unexpected obstacle appears, or when a key does not work in a familiar lock).

Much of what we know and that comes to mind is not the result of any deliberate search. In recent years there has been increasing research activity related to so-called implicit (as contrasted with explicit) memory.[12] For example, many amnesic patients cannot consciously recall events that may come to mind nondeliberately (implictly) when appropriately prompted. The question of the power of unconscious processes still needs further work, and recently, there has been some disagreement over the reality of an implicit–explicit distinction with respect to learning: Can new knowledge be acquired implicitly or can declarative knowledge be acquired only consciously?[13]

Another function of conscious construction is that it brings two or more previously separate mental contents into direct juxtaposition. The phenomenal experience of choice seems to involve exactly such an occurrence. We usually do not refer to a choice unless there is a "conscious" choice between two or more alternatives. The attribute of "choosing" is applied to a decision process regarding several items on a menu, several possible television programs, or two or more careers. Which particular choice we make is determined by complex unconscious mechanisms, but we need the prior selective function of consciousness to determine the events or objects on which these mechanisms operate. Selective consciousness generates the objects among which choices are made.

Consciousness also generates the redistribution of activations; choice mechanisms operate on the basis of new values of schemas and structures that have been produced (activated) in the conscious state. The mechanisms that select certain actions among alternatives are not themselves conscious, but the range of objects among which one chooses is created consciously, thus contributing to the appearance of free choices and operations of the will. What consciousness does entail is running through potential actions and choices, the coexistence of alternative outcomes, and the changing activations of currently active schemas.[14]

People also develop representations *about* the representation of their thoughts and actions. One of the important distinctions that needs to be kept in mind in studying consciousness is that between the underlying schemas that construct thought and actions, on the one hand, and the cognitive structures that are glosses on them. These, best called *secondary structures,* describe and activate the underlying schemata. They are particularly evident with respect to action systems, which themselves often remain permanently inaccessible to consciousness. What is available to consciousness are these frequent secondary schemas about our thoughts and actions, represented in part in mental models, and in personal theories and beliefs. Secondary structures are also a shorthand way of describing or activating complex structures that are often unconscious in detail or as a whole. Commands are an example of such secondary referents; "Stir that sauce!" "Drive the car, please," or "Deal the cards." Other instances include so-called metacognitive statements, such as "I know how to do multiplication," "I can recite all the presidents of the United States," or "I make an excellent omelet." And even statements such as "He is stubborn," "I believe in God," and "That is unjust" are shorthand ways to describe complex and temporarily unconscious sets of schemas. Again, as I noted earlier, these descriptions of secondary structures are usually verbal because that is what is available for efficient communication. In fact, these secondary structures are usually complex concepts, and their verbal descriptions only refer to them.

The major sources of calling for a particular construction are current tasks and contexts, intentions, and needs. Just as current perceptions are the result of both external evidence and internal processes, so is consciousness in general determined by activated higher order structures as well as by the evidence from the environment. Structures that represent intentions and interpretations of situational requirements depend on prior evaluations, on activations of situational identifications and interpretations, and on current needs and goals. In the normal course of events, the more abstract and general structures define what we are doing, what we want to do, and what we need to do. When task and intention are narrowed down to particulars, less general and more specific schemas, intentions, and expectations determine conscious constructions.

I now turn to specific mechanisms and functions of consciousness—first to its limited capacity and seriality and then to its feedback function.

Limited Capacity and the Utility of
Consciousness as a Serial Device

The fact that our conscious perceptual capacity is limited to a small number of bounded events or objects has been known for centuries, and it became a central part of modern information-processing approaches when George Miller described it as an important limitation of human thought.[15] The limited capacity of consciousness serves to reduce the "blooming confusion" that the physical world potentially presents to the organism. The world is full of events that our senses do not and cannot register, and just as our sensory organ systems radically reduce and categorize this world of physical stimuli to the functional stimuli that are in fact registered, the conscious process further reduces the available information to a small, manageable, and serial subset. We "think" (are conscious of) in a serial manner; we cannot think about more than one coherent set of perceptions or ideas at a time. And "coherence" refers to the "limited" aspect of our conscious thought—which cannot contain more than five or six constituents.

The question arises of whether we are really "conscious" of five or six discrete events. I have argued that only one "view," one conscious construction, of a particular event is generated at any one time and that such a construction responds to intentions and tasks of the moment. If we consider consciousness as an integrated construction of the available, relevant evidence, a construction that is phenomenally "whole," then the limitation to a certain number of items or objects or events or chunks refers to the limitation of these elements *within* and *by* the structures that make up the conscious experience. The schemas that are represented in the conscious construction are necessarily restricted to a certain number of features or relations. Cognitive "chunks" (organized clusters of knowledge) operate as units of such constructed experience. However, only a limited number of such chunks make up the current conscious experience. For example, as I look out my window, I am aware of the presence of trees and roads and people, a limited number of individual organized schemas that make up "the view." I may switch my attention and reconstruct my conscious experience to focus on one of these events and note that some of the people are on bicycles, others are walking, some are female, and some male. Switching attention again, I see a friend and note that he is limping, carrying a briefcase, and talking with a person walking next to him. At that point, the trees, the people on bicycles, etc., are not part of my current consciousness any more. In each case, a new experiential whole enters the conscious state and consists of new and different organized chunks.

The organized (and limited) nature of consciousness is illustrated by the fact that one is never conscious of totally unrelated events. In the local park I may be conscious of some children playing, or of parents with baby carriages, or of children and parents interacting, or of some people playing chess; but the conscious representation that includes a child, a chess player, a father, and a baby carriage not in some meaningful action or relationship is unlikely. In arguing that the limited capacity of consciousness is represented by the number of events that can be organized within a single constructed conscious experience, I respond to the intuitively appealing notion that we are both aware of some unitary "scene" and have available within it a limited number of constituent chunks.[16]

The selective nature of consciousness, the gate between external information and internal representations, constrains what we "learn"—what new concatenations of events are retained for future use. And this gate is at the same time the entry to the parallel mechanisms active "below" consciousness. As I have noted, it is a highly limited system, and it is also serial. It is these limitations that slow down the multiplicity of parallel mechanisms and, most important perhaps, that provide selectivity.[17] The activation of selected unconscious processes by current conscious constructions selects specific events and objects. At the same time, consciousness acts as a gate so that only information that is currently relevant (i.e., fits the regnant conscious constructions) is encoded for storage and future constructions.

To illustrate the importance of limited, serial conscious representations, imagine consciousness as it is, behaving as yours does, but with only one exception—namely, its seriality. Imagine consciousness as a parallel machine that permits everything currently relevant (or unconsciously active) to come to consciousness all at once. You would be overwhelmed by thoughts, potential choices, feelings, and attitudes of comparable "strength" and relevance. As you read a book, all the characters and their implications would cascade in your mental life. Consider the story of Lord Nelson and Lady Hamilton: As you read of one their trysts you would also be aware and conscious of his victory at Trafalgar, his defeats in the Mediterranean, his antirepublicanism, and his narcissism and of her eventual obesity, her Lancashire beginnings, her lovers, and her husband's interest in classical vases and volcanos, and . . .[18] A huge mishmash of associations and ideas would envelop you, and that discounts simple environmental events such as the chair you are sitting on, the lamp that illumines your book, and so forth. This is a "humanly" impossible situation. All of this would come in simultaneous snippets, still constrained by the limited capacity of the machine. In this account I have not relaxed the constraint of limited capacity. To relax that restriction too, to permit all unconscious content to become conscious, might strain the capacity of the reader to suspend disbelief. But wait just one more moment. Would that consciousness not remind you of a consciousness discussed in some other place? Is that not a description of God—aware of all that all of "his children" (including the merest sparrow) ever do and think? Can one really move that easily from humanity to deity—just by suspending seriality, limited capacity, and the current relevance of consciousness?

The Feedback Function of Consciousness

One of the important characteristics of consciousness is that it can have effects on *subsequent* mental events and (physical) actions. The feedback assumption contrasts with the view that consciousness cannot have any causal effects. Conscious phenomena appear to occur *after* the event that they register[19] and seem to be causally inert—that is, "[c]*onsciousness is not good for anything*"[20] In contrast, the feedback assumption asserts the causal utility of conscious events, as well as their effect on subsequent activations of the consciously represented events.

The feedback assumption states that the alternatives, choices, or competing hypotheses that have been represented in consciousness will receive additional activation and thus will be enhanced.[21] Given the capacity limitation of consciousness, combined with the intentional selection of conscious states, few preconscious candidates for actions and thoughts will achieve this additional, consciousness-mediated activation.[22] What structures are most likely to be available for such additional activation? It will be those preconscious structures that are most responsive to current demands and intentions. Whatever structures are used for a current conscious construction will receive additional activation, and they will be those most relevant to current concerns. Feedback and its resultant activation form a continuous process, and what is currently conscious will be so because of prior feedback. Alternatives that may have been candidates for conscious thought or action but were not previously activated will be relegated to a relatively lower probability of additional activation and therefore are less likely to be available on subsequent occasions.

The evidence for this general effect is derived from the vast amount of current research showing that the sheer frequency of activation affects subsequent accessibility of thought and action, whether in the area of perceptual priming, recognition memory, preserved amnesic functions, or decision making.[23] The proposal extends such activations to internally generated events and, in particular, to the momentary states of consciousness constructed to satisfy internal and external conditions. Thus, just as reading a sentence produces activation of the underlying schemata, so does (conscious) thinking of that sentence or its gist activate these structures. In the former case, what is activated depends on what the world presents to us; in the latter the activation is determined and limited by the conscious construction. But for the feedback function to make sense, we must assume that the constructive function of consciousness that selects appropriate mental contents is also operating. This hypothesis of selective and limited activation of situationally relevant structures requires no homunculuslike function for consciousness in which some independent agency controls, selects, and directs thoughts and actions that have been made available in consciousness.

The proposal can easily be expanded to account for some of the phenomena of human problem solving. I assume that activation is necessary but not sufficient for conscious construction and that activation depends in part on prior conscious constructions. The search for solutions to problems and the search for memorial targets (as in recall) typically have a conscious counterpart, frequently expressed in introspective protocols. What appear in consciousness in these tasks are exactly those points in the course of the search when steps toward the solution have been taken and a choice point has been reached at which the immediate next steps are not obvious. At that point the current state of the world is reflected in consciousness. That state reflects the progress toward the goal as well as some of the possible steps that could be taken next. A conscious state is constructed that reflects those aspects of the current search that do (partially and often inadequately) respond to the goal of the search. Consciousness at these points depicts waystations toward solutions and serves to restrict and focus subsequent pathways by selectively activating those that are currently within the conscious construction. Preconscious structures that construct consciousness at the time of impasse, delay, or interruption receive additional activation, as do those still-unconscious structures linked with them. The result is a directional flow of activation that would not have happened without the extra boost derived from the conscious state.

Consider the following example: As I tried to remember when and with whom I undertook a particular unpublished experimental study, I first recalled the names of some students and colleagues who were interested in that problem. None of the names seemed to satisfy the goal of my search, and I recalled next that one of these people had a passing interest in that experiment. I then recalled that this person had an office mate whose name escaped me for the moment, and I tried to reconstruct the physical layout of our labs and offices. I recovered an image of that office mate, then remembered her name, and immediately remembered that, while not directly involved in the project, she had offered to conduct the experiment with me. The conscious waystations and how they might have redistributed activations appear in this account. Whenever I reached an impasse, conscious contents focused on the dilemma and provided new activations that led to new conscious states, and so forth, until a solution was reached. For example, one waystation included a representation of a selected set of students and colleagues; another one appeared as a result of the activation of this set together with the prior activation of the particular experiment and produced a narrower conscious focus of possible candidates.

In my discussion up till now I have concentrated on complex conscious contents, primarily those that seem to exhibit planning and the usual meanings of cogitation. These instances usually involve reflective consciousness—conscious contents about other conscious contents. What is usually involved is the dissection or unpacking of a consciously represented experience. This recursive aspect of consciousness seems to be a particular characteristic of rather complex organisms such as human beings.

In contrast to other slow and serial (but much less limited) representations, as in motor skills and some other automatic sequences, consciousness permits the recursive representation of conscious contents and the analysis and dissection of their constituents. Finally, these complex conscious representations need to be contrasted with conscious representations such as the awareness of simple stimulus events such as colors, weights, sizes, etc. There is an obvious difference between perceiving the color red and consciously solving a mathematical puzzle. These cases produce similar phenomenal experiences but may in fact be the product of developmentally (and evolutionarily) different processes.

It is equally important to say what, under this view, consciousness is *not*. It is not a homunculus with its own motives and mechanisms to be regressively explained. Various presentations by members of the cognitive fold have explicitly or implicitly assigned an "executive" function to consciousness. Apart from the predilection of contemporary theoretical language to speak in corporate metaphors, the "executive" function of consciousness is apparent—not real. To stay within the metaphor, consciousness acts more like an executive secretary who decides what is and what is not (and in what form) relevant to current concerns. And this gating function is performed by a well-trained robot who does only what the job description specifies—it involves no independent judgment.

I also find some philosophical discussions peripheral to the central question of consciousness—that is, how it functions in an information-processing system. Some such speculations have failed to make adequate contact with the existing psychological literature,[24] but most seem to invoke fanciful distinctions that make little difference to a causal account of the role of consciousness.

I summarize this discussion by suggesting that some minimal number of assumptions and requirements might do justice to the known functions of consciousness. We may be able to entertain an understanding of conscious phenomena by considering only three basic functions of consciousness:

1. The selective/constructive representation of unconscious structures/ schemas.
2. The conversion from a parallel and vast unconscious to a serial and limited conscious representation.
3. The selective activation (priming) by conscious representations that changes the unconscious landscape by producing new privileged structures.

Whether or not these minimalist assumptions are adequate to handle the most obvious or inferred functions of consciousness will depend on future research and insights. I now turn to specific constructions of consciousness that characterize our daily lives.

Constructions of Consciousness

The Construction of Consciousness in Daily Life

I have argued that current conscious contents are responsive to the immediate history of the individual as well as to current needs and demands.[25] To recapitulate: We start with the (unconscious) schemas that represent current mental life. Schemas are dispositional mental structures that are constructed or assembled out of distributed features. The unconscious mind is not a library of static schemas, but rather an assemblage of features and attributes that produce appropriate representations. Currently available information constructs (out of schemas and features of previously developed schemas) representations that respond both to the immediate information and to regularities generated by past experience and events. Evidence (occurrences) from both extra- and intrapsychic sources activates (often more than one) relevant schemas as well as specific memories. More abstract and generic schemas will be activated by spreading activation. These assemblages of features and temporarily activated schemas provide the building blocks for conscious representations. I will experience (be conscious of) whatever is consistent with my immediate preceding history as well as with currently impinging events. The most important schemas that determine current conscious contents are those that represent the demands and requirements of the current situation.[26]

The current situation activates and constrains schemas (hypotheses) of possible actions, scenes, and occurrences in terms of one's past experiences. In other words, current conscious contents reflect past habits and knowledge in addition to, and often instead of, the representations of the "real world."

One of the best examples that conscious contents respond not merely to "veridical" representations is illustrated by research that has shown that conscious reconstructions of previous events reflect not just what "actually happened," but also respond to variables and structures of which we are not conscious and which distort "veridicality."[27] Distortions (constructions!) of conscious memory, as, for example, in eyewitness testimony, provide many instances of this process. Vibration-induced illusions (sensory misinformation) of limb motion produce novel but "sensible" apparent body configurations, so that, for example, vibrating the biceps of the arm while one's finger rests on the nose produces the experience of an elongated nose.[28] Similarly, misleading information about one's hand movements apparently requires and produces the experience of involuntary hand movements.[29]

Until now my discussion of conscious constructions has had an all-or-none flavor—either a particular state is conscious or it is not. However, there are potential and partially realized actions and thoughts that seem to hover on the edge of consciousness. I wish to talk about these in terms of a state-of-the-world knowledge that is primarily dependent on schemas and thoughts which, as a result of both our intercourse with the external world and our own mental activities, continue to stay active and quickly available.[30] Although relatively little is known about the course of activation—how schema activation decays and how that decay differs for different kinds of structures—we can safely assume that, once activated, many mental structures maintain an increased level of activation for some reasonable length of time.

Our ongoing experience of the world continuously updates the activation of schemas and other cognitive structures, both specific and general. Starting with one's morning activities, and probably with residues from the previous night's dreams, schemas relevant to eating breakfast, getting dressed, planning the day's activities, and interacting with one's family, all are activated and probably stay in a state of activation for some time. As the day proceeds, more and more previously unconscious schemas are activated, though not necessarily used in conscious constructions. The set of activated mental structures defines our state-of-the-world knowledge. Evidence from events in the world, thoughts about today's and previous activities—in short, all we do and think—affects the structures and schemas of our unconscious and activates them. These directly or indirectly activated representations may subsequently be easily and often automatically brought into consciousness. Constructions involving currently active schemas come to mind seemingly without effort; they have the phenomenal appearance of hovering close to consciousness. No deliberation is needed to determine who I am, where I am, or what I am doing, even though I may not have been consciously thinking about any of this knowledge just prior to its realization. These preconscious structures determine our immediate expectations as well as our current state. These expectations (and their underlying schemas) are continuously updated. As I sit in my office, I "know" that I am the only person there, until a colleague comes in and asks to look at some journals. I return to my work, without consciously thinking about her, but I now "know" that there is somebody else there and I will not be surprised to hear her ask a question. These current activations determine the current "meaning" of the world around us; they are representative not only of our immediate experience with the world but also of wider themes. Both worldly and internal interactions activate specific and concrete schemas on the one hand and more general categories and schemas on the other. A particular interaction may have activated a representation of cooperation and helping, which will in turn exert its influence on subsequent interactions. Solitary musings about some of yesterday's events or about a long-past experience will also make available those structures for interactions with current events. Finally, as some psychologists seem to forget at times, action and thought, and the activation of the relevant cognitive structures, are both multi- and overdetermined. Conscious contents will be determined by a large variety of current and past experiences that have, for one reason or another, produced relevant activations at the preconscious level.

Emotional experiences are also actually or potentially with us in the same state-of-the-world activations. Changes in the contents of consciousness are, to a large extent, due to failures of expectations, changes in the currently experienced state of the world, and I shall discuss in the next chapter how and why such changes may generate some emotional states that are likely to be a constant feature of an active mind. These "emotional" states are relatively weak and transient. Absent important and drastic changes in our world, these "little emotions" color our ongoing life; they do not usually dominate it.

There is one aspect of consciousness that is rarely considered—the relationship between normal, everyday consciousness and esoteric and altered states of consciousness. I have discussed Zen, Sufi, and other meditative practices elsewhere,[31] but others have generally ignored the phenomena practiced and discovered by these, to us, esoteric approaches and methods. In particular, I have suggested that the ability of some of the practitioners of meditative arts to focus and meditate on some single aspect of their environ or experience seems to produce "conscious stopping," similar to a frame-freezing (movielike) experience. Recent neurophysiological speculations and evidence by Llinàs and his associates have also suggested that consciousness may be a rapid succession of discrete states.[32]

It is the hallmark of sane "rational" adults that they are conscious of a world that is consistent, that is "real," and that is experienced the same way by other people. But there is another frequent human activity that is relatively unconstrained by reality, yet is conscious—namely, our nightly dreams.

Dreams, Reality, and Consciousness

In contrast to everyday life, in dreams possible constructions are only partially constrained by current reality and by the lawfulness of the external world. Yet dreams are highly structured; they are not random events. They present a structured mixture of real-world events, current events (sensory events in the environment of the dreamer), contemporary preoccupations, and ancient themes. They may be weird and novel, but they are meaningful. What they are not is dependent on the imperatives and continuity of the real world—inhabited by physically and socially "possible" problems and situations. In the waking world our conscious experience is historically bound, dependent on context and possible historical sequences. In contrast, dreams do not depend on current sensory activations; they are constructed out of previous activations. Similar arguments have been made that during dreams (REM sleep) "the brain is isolated from its normal input and output channels."[33] The brain/mind is focused in the waking state on the linear unfolding of plot and time, whereas during REM sleep the brain/mind cannot maintain its orientational focus.[34] The leftovers of our daily lives are both abstract themes—our preoccupations and our generic view of the world—as well as concrete and specific activated schemas of events and objects encountered. These active schemas are initially not organized with respect to each other, they are, in that sense, the random detritus of our daily experiences and thoughts. Without the structure of the real world, they are free floating. They are "free" to find accommodating higher-order structures. These may combine quite separate unrelated thoughts about events, about happy or unhappy occurrences, but since there are few real-world constraints, they may be combined into sequences and categories by activating any higher-order schemas to which they may be relevant.

It is in this fashion that abstract (and unconscious) preoccupations and "complexes" may find their expression in the consciousness of dreams. It is what Freud has called the "residue" of daily life that produces some of the actors and events, whereas the scenario is free to be constructed by otherwise quiescent higher-order schemas.[35] These themes of dreams may be activated by events of the preceding days, or they may be activated simply because a reasonable number of their features have been left over as residues from the days before. I should note that dream theories that concentrate only on the residues in dreams fail to account for the obviously organized nature of dream sequences—however bizarre these might be. In contrast to mere residue theories, Hobson's activation-synthesis hypothesis of dreaming supposes that, apart from aminergic neurons, "the rest of the system is buzzing phrenetically, especially during REM sleep."[36] Such additional activations provide ample material to construct dreams and to be creative and to generate solutions to old and new problems.[37] This view is not discrepant with some modern as well as more ancient views about the biological function of dreams (in modern times, specifically REM dreams), which are seen as cleaning up unnecessary, unwanted, and irrelevant leftovers from daily experiences. However, these views of dreams as "garbage collecting" fail to account for their organized character.[38]

In short, dreams are an excellent example of the constructive nature of consciousness: they are constructed out of a large variety of mental contents, either directly activated or activated by a wide-ranging process of spreading activation, and they are organized by existing mental structures.

Memory and Consciousness

Human memory is an excellent proving ground for notions about the functions of consciousness. The common understanding considers as memorial those events that reach a conscious state, whether access is achieved consciously or unconsciously. I have previously mentioned the distinction between implicit and explicit processes—that is, between memories that are brought to mind deliberately (conscious/searching) or nondeliberately (unconscious/automatic). The former have the appearance of being voluntary memories, the latter of being involuntary memories.

Anterograde amnesia, the inability to learn or remember new experiences, provides some relevant demonstrations of the functions of consciousness. An appeal to the functions of consciousness provides the theoretical formulation for an understanding of the memorial disability that we find in anterograde amnesia. For purposes of this discussion, I restrict myself to the kind of phenomena typically found in patients suffering from Korsakoff's syndrome. These amnesic patients suffer from an inability to use the conscious function of creating new concatenations, new contexts for old knowledge, or conscious access routes to information that is not available automatically. In short, the patient has a deficiency in conscious functioning that prevents the storage of new information or the retrieval of previously stored material, unless it can be done automatically and without the intervention of the conscious apparatus. Most prior discussions of amnesia have stressed what it is that is spared in amnesic patients.[39] An appeal to conscious functions describes both what is impaired as well as what is spared. What is impaired is the ability to construct new conscious contents or to retrieve recently encoded information.

What is it Like Not to Be Conscious?

One could speculate what human life would be like without consciousness, but it would be difficult to find any evidence for such speculations. However, there is some evidence on the limitation of consciousness which provides a partial insight. Various reports on patients with very dense amnesias have described the pathologies of consciousness which I wish to invoke. Patients describe themselves as being "prisoners of the present," as experiencing only "momentary consciousness." They appear to construct conscious contents only as a function of current activations, when there is "bottom-up" activation, or in terms of pretraumatic automatically accessed memories. What they cannot do is use or construct novel conscious posttraumatic contents that are stored and retrieved as a unit. Tulving has presented a case that illustrates this conscious inability, this failure to construct conscious contents.[40] One extract of a patient interview has the interviewer ask what the patient will be "doing tomorrow." The answer is that the patient "doesn't know" and his mind is "blank" when trying to think about it. He is living in a "permanent present" and cannot remember particular episodes in the past or construct possible futures. The patient is unable to construct conscious contents about the past or present—concatenations of new or imagined events—or between the present and the past. On the other hand, knowledge that does not require novel constructions is generally unimpaired; the "semantic" memory is intact. Thus such a patient can perform his daily routines and converse intelligently.

One could imagine that "living in the present" may well describe the experiential world of some animals. In fact, some descriptions of the conscious limitations in severely impaired amnesic patients include statements such as "I am no more than an animal." The less complex animal mind may construct only a momentary consciousness, depending on bottom-up current evidence rather than on top-down constructed "intentions." We reflect this state of affairs when we talk of some lower animals as being "stimulus driven."

If living in the present characterizes other organisms, such as some lower animals and very young infants, are there other mechanisms that perform some of the functions of consciousness? Examples of such precursors of consciousness are the vicarious trials and errors (VTEs) observed in the choice-point behavior of lower animals.[41] VTEs consist of the visual sampling of possible choices, sometimes including short forays to one or the other side of the choice point, before the actual choice occurs. Such "sampling" increases the relevant activations of underlying representations when there is no representation in consciousness to perform that function of selective activation. The same kind of behavioral, in contrast to mental, sampling can also be observed in adult humans (e.g., under stress) and in infants. Let me suggest that VTEs are frequently no more than the external signs of spatio-temporal "attention." Such an attentional phenomenon frequently gives rise to the attribution of intentionality and consciousness to the attending organism. Attention to objects and space-time events is by itself not evidence for a conscious organism.

Some Speculations About the Evolution of Consciousness

As far as the evolution of consciousness is concerned, I suggest that the need for and selection of some monitoring system occurred when the complexity of tasks to be accomplished and information to be monitored became sufficiently demanding. The secondary "conscious" system may have emerged with respect to only one or some of the simpler functions of consciousness. Subsequently, and in a homologous fashion, the system was adopted for other functions and, over the millennia, developed into a complex system of its own. Gould has made a similar argument that important (conscious) cognitive functions are not an orderly consequence of biological improvement, but rather the fortuitous consequence of neural complexity evolved for other reasons. Specifically, "as a result of larger size, and the neural density and connectivity thus imparted, human brains could perform an immense range of functions quite unrelated to the original reasons for increase in bulk."[42] In regard to some of the important functions of consciousness, how does consciousness contribute to the procreative advantages of the conscious organism? Presumably, the serial and limited capacity aspects of the conscious system contribute to probabilities of survival and selection by making decisions both quicker and more relevant to the current situation. An organism that is not overwhelmed by too many alternatives, that is able to consider the relevant ones in an orderly fashion one at a time, is more likely to survive (and procreate) than one that may be panicked by the rush of possible actions.

We still develop consciousness—ontogenetically—as it is needed and made possible by the requisite underlying structures. The most striking representation in consciousness is, of course, the representation of the self, of the structures and schemas that represent knowledge of the persons that we are, our needs and wishes, fears and hopes. Again, I argue that this ability to be aware of ourselves is not a separate characteristic developed or given at some blinding point in time, but rather it is one other—and overwhelmingly persuasive—result of the development of consciousness.

This point of view does not view consciousness as an all-or-none phenomenon—either you have "it" or you don't. Instead, it emerges in response to the demands of processing in special domains and in different domains at different times. Consciousness may or may not be a function exclusive to human beings, and, more important, it did not suddenly or recently come into existence. Such a view is in direct contrast to that of philosophers and psychologists who give consciousness a uniquely human flavor, circumscribed by language and human mentality. Instead of thinking of consciousness as a single attribute or characteristic, I prefer to speak of different mechanisms that produce similar phenomenal experiences. These may range from the phenomenal experience of a particular quality or stimulus (such as red, heavy, or even beautiful) to complex reflective cogitations (such as planning a party or analyzing a recent unpleasant interaction with a colleague).

One of the consequences of such a position is that complex consciousness, at least, is not a given characteristic of an individual person or animal, but rather the individual has the potential to construct a conscious content only if some conscious representational mechanism exists for an important event or some higher-order structure is available that organizes the relevant experience and permits reorganization or reflection. Thus, people are not and cannot be conscious of just anything in their environments or in their psychic armamentarium, but they may become conscious of specific kinds of events in their internal or external environment. Which kinds of events that includes in the case of complex conscious contents depends on the prior experience one has had with the phenomena and experiences in question. This accounts for painters and musicians, for example, being able to perceive (consciously) aspects of art and music that the untrained eye and ear do not perceive. Prior experience with the constituents is necessary, as is some initial construction of, need for, and knowledge about relevant higher-order structures. In other words, the development of consciousness about certain objects and events may be a slow and cumulative process.

This kind of approach may also be useful in understanding the development of consciousness in the young child. Thus, we would not expect that an infant suddenly becomes conscious of its entire world, but rather that the ability to construct conscious contents will depend on its developing experience with its surroundings. For example, one would expect the child to form early consciousnesses about its caretakers and about hunger management. On the other hand, consciousness about language would take quite some time until specific initial structures are developed which then may become the constituents of higher-order structures that construct conscious knowledge about the use and structure of language. Thus, the metalinguistic abilities of children—what it is that the child knows about language that the spider does not know about its web—will emerge with the development and mastery of language.[43] A somewhat different view of the developmental origin of conscious processing concludes that the cognitive skills exhibited by infants as young as six months require some rudimentary conscious representations.[44] If that persuasive argument is correct, my account has to be amended to include such an early appearance of consciousness. The further enhancement and expansion of conscious constructions, however, could still proceed as described.

One can take an additional speculative leap and consider the problem of animal consciousness. It may not be the case that mammals are either conscious or not, but rather that the ability to construct conscious contents has developed within the mammalian groups. The richness and extent of consciousness depends on the richness and extent of underlying cognitive structures that encode and structure knowledge of the world. Cognitively sophisticated animals, such as the primates, will have a more extensive conscious apparatus than less sophisticated animals such as rodents or canines. The question may be not whether the animal is conscious, but what limited number of events and objects it can be conscious of.

The selective nature of consciousness, the gate between external evidence and internal representations, constrains what we "learn"—what new concatenations of events are retained for future use. Since contiguity—the mere contiguous appearance of information—appears to often be a sufficient condition for the acquisition of new knowledge, complex organisms whose attentional and sensory mechanisms are receptive to a wide variety of events might need to know about events and concatenations that are not relevant to the individual's current needs and concerns. If, in complex systems (complex minds), the secondary conscious system can segregate what is or is not self-relevant, then the organism is not overwhelmed by irrelevant knowledge. And if conscious contents are indeed selected by current relevance, needs, intentions, and salience, then new knowledge that is accepted and encoded within the conscious realm will be selectively relevant to the organism. Such selectivity is not needed for lower animals or simple minds where everything (or nothing) is self-relevant. In fact, one might speculate that for minds that are in the process of being constructed or that are minimally responsive to environments, it is desirable to have new knowledge built up automatically (i.e., without the interface of consciousness). Thus it appears that in the very young human infant, where few conscious gates have been established, much new knowledge is acquired procedurally—that is, unconsciously and often unselectively.[45] It is such knowledge that amnesic patients are able to acquire as well. Conscious processing, on the other hand, produces new knowledge that is independent of current stimulus-response contingencies. Such new learning responds to goals and concerns that are not exclusively a function of current situations and contexts. It is distinguished from new information that is acquired automatically, as in the case of many perceptual and motor skills and other simple "associations" that are stimulus driven rather than the result of deliberate knowledge acquisition.

If consciousness is a function of complexity, then its physiological embodiment is not likely to be found in any particular physiological units (cell, neuron, column, assembly), but is to be seen as an emergent characteristic of the system. A similar argument was made by Alfred North Whitehead when he noted that electrons within living bodies behave as they do as a function of the plan (structure) of the body and not because of any inherent characteristic. Complex arrangements of atoms in molecules are not "inherent" in the atoms themselves, and even such complex phenomena as the "handedness" of quartz crystals are the result of a chemical process; they are not "embryonically present in the atoms."[46] Nor is consciousness inherent in the neurons. Qualitative changes occur as a result of quantitative ones (as in the magnetic property of iron as it is cooled). Complex organisms and complex organs (such as the brain) can produce qualitatively different characteristics (such as consciousness) that are inherent neither in less complex organizations (e.g., in some other animals) nor in their constituents (neurons). Explanations at the psychological level address a different set of laws than do any possible physiological explanations of the same phenomena. Explanations at a "lower" level would, in fact, conceal the laws and relationships at the "higher" level.[47]

Finally, it has become obligatory in a discussion of consciousness to ask how a physical brain can generate mental qualia such as perceptions of color. The question has produced many premature explications, however ingenious some of them are.[48] I assume that, eventually we will understand this puzzle, just as we are beginning to understand how some emotional "feelings" may be generated. But the question about qualia requires a far greater knowledge of neurophysiology than we have at present and a healthy agnosticism—a resounding "I don't know"—might well be the best current position. We do not know, and we might know sometime in the future, but is the question really different from any other island of human ignorance? If we have truly abandoned Cartesian dual-ism, then one may entertain the question of how the brain "does" qualia or con-sciousness, seen as just another thing that it does.

I now turn to a discussion of emotional experience, so well represented in consciousness, and so often misunderstood.

Emotion

∞

Poetry . . . takes its origin from emotion recollected in tranquillity.
William Wordsworth

THE INCLUSION OF emotion in a book on human nature is unusual; the topic is not frequently considered to be specific to the nature of humanity. The reason for that exclusion is found in part in the view that human emotion is not that different from animal emotion. The basis for this belief is the emphasis on so-called emotional behavior—angry cats are said to look like angry humans, fearful rats are like humans afraid, and anxious dogs seem to be subject to human angst. Be that as it may, my major concern is not with behavior but with emotional experience, which drives much of our daily interaction and which colors our poetry, art, and music—if often only in recollection. Humans without hot emotions would be very different animals, as would humans without emotional consciousness. Thus, emotion becomes a central part of our nature.

There is another reason for including emotion here. I stress a proximate analysis of human emotion; it is an analysis that requires an unpacking of emotional phenomena as well as a demythologizing of emotional states. This discussion gives me an opportunity to demonstrate how complex human experiences, such as emotion, can be understood in terms of more simple constituents.[1]

Emotions: Approaches, Problems, and Myths

Emotions are the subject of many myths about humankind, and emotions are the myths whereby we flavor our existence and our beliefs. The challenge to understand emotions has created a flurry of theoretical attempts by philosophers and psychologists, just as poets and novelists have accepted emotions as overweening and inscrutably persistent. Emotions seem to come on us unbidden and often unexpectedly, we despair at controlling them, and they undermine our belief in the rationality of human existence. The categories of emotions, though they differ somewhat from culture to culture, have been established in the common language and have assumed a life of their own, a reification of families of usage that long ago installed such categories as fear or rage or anxiety or love as if they were cleanly delimited events or even palpable objects.[2]

The myths of emotion have arisen out of these common experiences, and they have persisted in the psychology of emotion. They encompass the sense that emotions are unitary, unanalyzable "packages" of experience, that there are some few basic emotions, and that they find expression in the human face. All three of these myths are the consequence of a belief in a single underlying disposition or program, associated with each of the basic emotions. These views are often held with fervor and passion; experience seems to allow nothing else.[3]

What Is An Emotion?

What is an emotion? was the question that William James posed over a hundred years ago and which has, on the one hand, had the beneficial effect of encouraging the study of "whatever it is," and, on the other hand, produced a search for an answer to a pseudo-question, or perhaps invited a confusion of "a semantic or metaphysical question with a scientific one."[4] As we know—and as I hope to show—different people answer the question differently, as behoves a well-used umbrella term from the natural language. Everybody wishes to unpack "emotion" idiosyncratically.

James, more than any one else, established the tradition of unpacking commonsense notions about the emotions, though in the process he misled several generations of psychologists into believing that his "What is an emotion?" admitted of an unequivocal answer.[5] James was primarily interested in the relationship between emotional feeling and bodily expression. He criticized the received knowledge of his day, which described emotion as a "mental affection" that "gives rise to the bodily expression." He noted that common experience suggests instead that our feeling of the bodily changes that follow the perception of some "exciting fact" *is* the emotion. Ten years later, James[6] started to turn to the problem of defining the "exciting facts," a problem central to cognitive accounts of emotion today but rarely, if ever, satisfactorily resolved.[7] It is often forgotten that James argued that if all feelings of bodily symptoms were abstracted from the felt emotion, all that would remain would be a "cold and neutral state of intellectual perception."[8] As an illustration, James noted that it would be impossible to think of the emotion of fear if "the feelings neither of quickened heart-beats nor of shallow breathing, neither of trembling lips nor of weakened limbs, neither of goose-flesh nor of visceral stirring, were present."[9] James was wrong in assigning the feelings of a particular emotion to a specific concatenation of visceral and muscular activities,[10] but his assertion about the centrality of visceral and muscular involvement has, unfortunately, been forgotten by most contemporary psychologists.

People seem to know full well, though they have difficulty putting into words, what emotions are, what it is to be emotional, what experiences qualify as emotions, and so forth. However, these agreements vary from language to language and from community to community.[11] Given that the emotions are established facts of everyday experience, it behooves us to determine what organizes the common language of emotion in the first place and then to find a reasonable theoretical account that provides a partial understanding of these language uses. Given the vagaries of the common language, it is, of course, useless to try to give a full account; the common language is neither exact nor universal. I follow James's lead by insisting that a reasonable account of "emotion" must, in the first place, concentrate on "hot" emotional experiences because it requires a separate account from all the other myriad uses that have been invoked and because any other approach leads to confusion and away from a clear understanding.

Is there a common core to the various uses and misuses of emotion language? Is there anything that is essential to the use of the term "emotion," some aspect that represents the core, without necessarily doing justice to all the nuances and implications of the concept? Lexicographers perform an important function for the social sciences; they circumvent the need for extensive surveys and interviews by distilling the meanings of our language. Their work is cumulative and, in general, responds to the nuances and the changing customs of the common language. The 1991 *Webster's New Collegiate Dictionary* defines *emotion* as "a psychic and physical reaction subjectively experienced as strong feeling and physiologically involving changes that prepare the body for immediate vigorous action," and *affect* is defined as "the conscious subjective aspect of an emotion considered apart from bodily changes." This traditional definition includes an approach to James's abstracted emotion. Presumably under the influence of the established wisdom since Aristotle, emotions are seen as having two components, a psychic and a physical one.

All science is, in the first instance, an extension of common sense and common knowledge. In the social sciences, and particularly in psychology, refined ("scientific") observations coexist to a large extent with common experience and common myth and lore. Emotions are taken to be unitary experiences, they occur in more or less well-defined categories (such as fear, love, hate), they are supposedly expressed in the face and body, and, being apparently not under rational, voluntary control, seem to be primitive and animal-like.

Unfortunately, psychological theory has often ignored some of the major insights of the folk psychology of emotion—such as the distinction between physical and psychic manifestations of emotion—and taken the remaining inconsistencies and uncertainties as license to build any kind of reasonable theoretical structure that seems to satisfy one or another set of frequently unjustified beliefs.

The multitude of contemporary theorists usually include autonomic arousal (the "fire in the belly") as an optional aspect of emotion but fail to address the question what the "exciting fact" may be—how we get to the hotness of the emotions. In the end, affect and emotion are treated as equivalent expressions, when even the common language does not permit us to say such things as "I am affectively in love with you," or "I found the situation affectively disturbing." Nowadays anything affective or evaluative is usually considered to be emotional. Statements of the order "I love chocolate ice cream," "My country right or wrong," "Isn't that a great car I bought," "I am sorry I hit my child"—all may be brought under the umbrella of this expanded notion of emotion as illustrating states of mind that represent such emotions as desire, patriotism, pride, and shame or embarrassment and extend to such various states as ennui, attraction, attitude, preference, relief, boredom, contentment, and so on. With that imperialist ambition for theories of emotion, is it surprising that there is so little overlap among these theories, that there are a large number of theorists who expound an even larger number of theories? These affective or evaluative statements deserve proper theoretical treatment, but by putting them into a grab bag called "emotion," no useful purpose is served. Central to all these approaches is the notion of value, whether specifically or implicitly approached, and I return to this topic in chapter 7. As I note later, values can be expressed in a variety of different modes, including verbal, expressive, facial, and metaphorical ones.

To start with a brief assertion of what I believe is a possible approach to "hot" emotion, in contrast to the theoretical confusion current in psychology, I limit my speculations about emotion to occasions that concatenate autonomic (gut) arousal and evaluative cognitions (or appraisals). This is the traditional view that goes back to Aristotle and Descartes, is emphasized by James, and is represented in modern times by Cannon and Schachter.[12] In recent years there has been a tendency to extend the notion of "emotion" to take in aspects, behavior, and topics that were not considered to be "emotions" prior to the late nineteenth century. And such hot emotions as lust (and some instances of love) are generally omitted from consideration.

Understanding Hot Emotions: Causes, Consequences, and Constructions

The history of psychology in the twentieth century has been one of analysis, an unpacking and reconstructing of the phenomenal experiences that represented psychological accounts up through the nineteenth century. In the process, unitary subjective experiences have been shown to be the result of specific underlying (usually unconscious) mechanisms and processes. Such a decomposition has been useful in vision, perception, audition, conscious motives and reasons, memory, and several other fields of psychology. One result of this effort has been a rapprochement with neurophysiology, since the specification of detailed mechanisms (the proximate causes of the subjective experiences) has made possible a more targeted search for underlying physiological processes.[13] But there have been several efforts at unpacking emotions that tried to explain them without recourse to the myths; these, however, were either ignored or circumvented.[14] An analytic, constructivist position asserts the coaction of two major components: autonomic (visceral) arousal and cognitive evaluations and analyses.[15] The former provide the hot, passionate background to emotion, the latter the specific content—the meaning of the emotion.

Before detailing my constructivist position, I turn to a major mechanism—the detection of discrepancy that provides the cause of arousal and the initiation of "excitation." It is in part an answer to James's search for the "exciting fact" and in part a contribution to an understanding of the evolutionary background of human emotion.[16]

Difference Detection: Discrepancies Are Important

One of the most important evolutionary inheritances which we share with other animals is the discrepancy response to the environment, the fact that we respond to states of the world around as "different" or "same." Any organism has a repertory of either innate or acquired responses and actions to specific aspects of their environment. Thus, we learn how to walk up stairs, respond to changes in gravity, act in the presence of painful sensations, learn what to do in a restaurant or in a garden, and so forth. But stairs need to have prescribed steps, gravity has a normal value, pains usually arise from unexpected situations, restaurants need tables and food, and gardens should have flowers. If any of these characteristics are different, it is useful to know that the setting has changed, and when they are as expected, it is useful to know that, too—that is, that they are "familiar." In addition, it is useful to have a mechanism that not only reacts automatically to changes in the environment but also prepares the organism for appropriate action or reaction. An organism that does not have a difference detector would be overwhelmed by a world which in fact often changes. Preparation and mechanisms for handling unexpected changes in the world are obvious adaptive mechanisms.

There is extensive evidence that our perceptual system reacts quickly and automatically to changes in the environment. The unexpected appearance of irrelevant objects or the absence of important objects in the environment immediately steer the perceptual analysis to the locus of unexpected changes of either commission or omission.[17] However, I want to stress a secondary automatic system that responds to change—the sympathetic nervous system (SNS). The SNS is concerned with energy mobilization; it is an emergency system that reacts to threat or danger by channeling blood supply to the muscles and brain, increasing heart rate and rate of metabolism, and raising the sugar content of the blood. Together with the parasympathetic system, which acts in energy conservation, the SNS constitutes the autonomic nervous system, which is responsible for the internal economy of the mammalian organism, which was represented in W. B. Cannon's concept of homeostasis. Cannon stated "[i]n order that the constancy of the internal environment may be assured, . . . every considerable change in the outer world . . . must be attended by a rectifying process in the hidden world of the organism. The chief agency of this rectifying process . . . is the sympathetic division of the autonomic system."[18] I assume that in addition to this passive adjustment, the activation of the SNS also alerts the organism to act on the environment, to change the environment to states where the "normal" internal environment is also restored. Thus, the function of SNS may be, in addition to its homeostatic one, to alert the organism that something in the world has changed, some action may need to be taken. It is, of course, in this sense also homeostatic—conserving the internal environment. Furthermore, the autonomic nervous system signals are relatively slow in arriving at the cortex, so that the SNS response is truly a secondary one, alerting the organism one to two seconds after the "cognitive" information has arrived and providing a fail-safe mechanism to make sure that attention is paid to the environment.[19] There is evidence that when no adequate response to or understanding of dangerous events has been acquired, emergency reactions occur "late"—that is, after the SNS signal has been registered. Thus, infants apparently do not respond to the pinprick of an injection, but to the later autonomic arousal.[20]

There is now enough evidence available to show that any disruption, interruption, or discrepancy—whether it occurs at the perceptual level or as a result of one's actions or behavior—does, in fact, inevitably produce SNS responses. We have shown it in simple situations of being interrupted in verbal sequences, in listening to simple stories, in playing computer games, etc.[21] Originally, I dubbed the phenomenon the "interruption of behavior," since it occurs when planned or ballistic ongoing behavior or action is prevented from being completed, but clearly discrepancies also occur when one's perceptual expectations are violated or when the world is simply not as expected.

The specific relevance of the SNS discrepancy response to emotional phenomena is that SNS arousal is part of the emotional experience and that discrepancies are the major occasions for emotions to occur. I stress that emotional states do not simply occur when the SNS is aroused—the cognitive or evaluative part is a necessary co-condition. In chapter 7, I discuss the relevance of discrepancies to problems of human values, as well as the converse of the difference detector—the basis for evaluating and experiencing the expected and the familiar.

The Construction of Emotion

The construction of emotion consists of the concatenation in consciousness of some cognitive evaluative schema together with the perception of visceral arousal. This conscious construction is, like most other experiences, a unitary experience, even though it may derive from separate and even independent schematic representations.

Among the possible analyses of common-language "emotions," I have focused on two such characteristics: the idea that emotions express some aspect of value, and the assertion that emotions are "hot"—they imply a gut reaction, a visceral response. These two aspects not only speak to the common usage, but also reflect a frequent conscious construction— that is, we experience value-laden hot emotions. One of the consequences of such a position is that it leads to the postulation of innumerable emotional states, no situational evaluation being quite the same from occasion to occasion. There are, of course, regularities in human thought and action that produce general categories of these constructions, categories that have family resemblances and overlap in the features that are selected for analysis and that create the representation of value (whether it is the simple dichotomy of good and bad, or the appreciation of beauty, or the perception of evil). These families of occasions and meanings construct the categories of emotions found in the natural language (and much of psychology). The commonalities found within these categories may vary from case to case. Sometimes they are based on the similarity of external conditions (as in the case of some fears and environmental threats), sometimes they share similar behaviors (as in the subjective feelings of fear related to flight), sometimes they arise from incipient behavior (as in hostility and destructive action), sometimes from hormonal and physiological reactions (as in the case of lust), and sometimes from purely cognitive evaluations (as judgments of helplessness eventuate in anxiety). In other words, the categories are fairly consistent internally but have little in common in terms of their origin or function. However different the sources of these various categories, they all tend to occur for most people and in most cultures. Their basis is in the interaction between our basic physical characteristics and the environments in which we live; the world is full of potential dangers, blocks to our goals, sexual objects, and so on. It is these commonalities that give rise to the appearance of fundamental or discrete emotions. However, these conglomerations are not haphazard collections; they are organized by their conditions and states, whether their source is behavioral, cognitive, or physiological. The source of their discreteness can be found in those conditions, rather than in a fundamental (biological) identity of their subjective feeling states. And even these "fundamental" emotional states require some analytic, "cognitive" processing. Emotions are frequently situation specific, and subjective emotional states, however one defines their source, need to be tied to some cognitive evaluations that "select" the appropriate emotion.

The problem of *cognitive evaluation* seems therefore common to all emotion theories. In recent years there has been much active search for the basis of these evaluative structures.[22] Basically, cognitive evaluations require some representation of value. What is the mental representation that gives rise to judgments and feelings of "good" or "bad" or of some affective nature in general? I suggest three such sources, to be discussed in detail in the following chapter:

1. Innate approach and withdrawal tendencies may be interpreted as communicating value. The avoidance of objects, the experience of pain (and the avoidance of pain-producing objects), and the taste of sweet substances are all examples of events that produce automatic approach and avoidance reactions. It is the secondary effects of these tendencies, when we observe that we are approaching or avoiding an object and conclude that we must therefore like or dislike it, that produce the judgments of positive and negative values.

2. Cultural, social, and idiosyncratic predication, which is the process by which objects, whether actually encountered or not, are assumed to have certain values as a result of social or personal learning experiences. Food preferences and aversions (to things such as frogs' legs and chocolate cake) are frequently acquired without any contact with the actual substances, as are likes and dislikes of people and groups of people. These predications, socially and idiosyncratically ac-

quired, produce judgments of value. Similarly constituted are culturally acquired aesthetic judgments of beauty—whether of people, landscapes, or paintings.

3. Structural value resides in the cognitive features of objects, and in the relations among these features, as in the appreciation of an object seen as beautiful or abhorrent as a function of a particular concatenation. Structural value differentiates patterns, rather than merely identifying objects, and arises out of our experience with objects and the analyses of their constituent features. One of the factors that influence judgments arising out of structural considerations is the frequency of encounter with objects and events.

If evaluative cognitions provide the quality of an emotional experience, then visceral activity—sympathetic nervous system (SNS) activity—provides its intensity and peculiar "emotional" feel. In keeping with the difference detection mechanism already discussed, visceral arousal usually follows the occurrence of some perceptual or cognitive discrepancy or the interruption or blocking of some ongoing action.[23] Such discrepancies and interruptions depend to a large extent on the organization of mental representations of thought and action. Within the purview of schema theory, these discrepancies occur when the expectations of some schema (whether determining thought or action) are violated. This is the case whether the violating event is worse or better than the expected one, and these discrepancies account for visceral arousal on both unhappy and joyful occasions. Most emotions follow such discrepancies, just because the discrepancy produces visceral arousal. And it is the combination of that arousal with an ongoing evaluative cognition that produces the subjective experience of an emotion. The effects of situational or life stress are excellent examples of unexpected events producing visceral arousal, negative evaluations, and emotional experiences.[24]

The invocation of discrepancies and arousal represents one attempt to understand when and why some emotions occur. To use arousal simply as an optional aspect of some complex emotional mechanism may tell us the "how" but not the "why" of hot emotions. The accompaniment of emotional states with the strong signal that visceral arousal provides also gives us some clues as to why highly emotional events are often easily remembered; they have additional cues and markers—the SNS perception. The co-occurrence of arousal with affective or value-laden states produces a very different set of phenomena than the mere registration of some affective state like a mood, an attitude, or a preference. Furthermore, there are affective states like tears that can be found in hot emotions, whereas others like contentment cannot. I can talk about being afraid or I can be intensely afraid, showing fear's cold and hot sides, respectively, but I cannot be passionately content or passionately bored. To mix the cognitive affective states with passionate, hot emotions creates confusion and misses an important point of human experience.

Finally, the construction of emotions requires conscious capacity. The very experience of emotion is, by definition, a conscious state and thus preempts limited capacity. As a result, emotional experiences frequently are not conducive to the full utilization of our cognitive apparatus, and thought may become simplified—that is, stereotyped and canalized—and will tend to revert to simpler modes of cognition and problem solving. The effect of strong emotions on thought and action is often deleterious—as in the extreme cases where panic states interfere even with life-preserving action. However, the effects are not necessarily intrusive and deleterious. The effects will depend, in part, on other mental contents that are activated by the emotional experience and that may become available for dealing with situations as well as mental events. The relationship of "emotions" to discrepancies and autonomic nervous system recruitment also points to their adaptive function; emotions occur at important times in the life of the organism and prepare it for more effective thought and action when, for example, focused attention is needed. The discrepancy–evaluation approach does not claim to deal with all the states and events that are called emotional at one time or another. All that the approach can claim is that it deals with a reasonably large subset of the conglomeration of emotional experiences.

Some Special Problems of the Emotions

Overdetermination

Emotions provide a prime example of the multi- and overdetermination of human thought and action. We do not live in our world with simple, single expectations, and similarly their violations (the discrepancies) are not unidimensional. Unconscious expectations of the state of the world not only encompass a large number of different aspects of our environment at the same time, but also involve conflicting expectations about the same event. One important example is found in many positive emotional states. The intensity of these states is often related to the fact that at the same time we expect (hope for) the positive outcome, we also "expect" the negative one. The intense delight of a student at receiving a high grade on an examination is probably in part related to her uncertainty—the mixture of expecting the good grade together with the possibility that she might get a lower grade. It is the latter expectation that is violated and provides additional arousal. As a result, together with the positive state engendered by the high grade, a positive emotional state is generated. Another student may receive the lower grade and his disappointment will be further potentiated if he had a positive expectation. Similar concatenations, often involving more than two activated schemas (and expectations), are found everyday in job situations, in close relationships, and even in rather mundane scenarios such as a visit to the dentist's office, eating a meal at a restaurant, or shopping at a supermarket.

The Ubiquity of Discrepancy

Discrepancy, as usually defined, seems to be ubiquitous. To some degree, all events are somewhat discrepant from what is expected; the world changes continuously. I would expect that there is some degree of arousal present in many, possibly most, day-to-day situations, and so is some degree of feeling. In fact, the cognitive-arousal position needs to account for the pervasiveness of moods and emotions, for the fact, that human beings are characterized by some feeling or mood state much of the time. It is the very pervasiveness of discrepancies that accounts for the continuous feeling states. However, the degree of discrepancy is usually slight and the amount of arousal is small, which accounts for some background "feelings." So-called true emotions, on the other hand, occur typically with high degrees of arousal and are associated with extreme discrepancies and interruptions.

The Nonspecificity of Arousal and the Experience of Lust

At the present state of evidence and research, it is not possible to identify specific patterns of autonomic arousal that are antecedent to and are causal in the experience of specific emotions. Experiential aspects of emotion seem to require some global autonomic event, and in fact our receptive systems are not adequate to differentiate among fine differences in different channels of the SNS mechanisms. There is, however, one class of experienced emotion that appears to be different—the experience of lust. Among the features that differentiate sexual lust are specific sets of prior autonomic events, including a brief parasympathetic episode prior to the onset of more massive sympathetic arousal. It is also the one emotional experience that has localized body experiences accompanying it—that is, genital sensory experience. Lust, however, is very different from the class of emotional experiences called "love," which appear to be no different in their accompanying SNS arousal from other emotional states.

More on the Positive and Negative Emotions

Both positive and negative emotions display complexities of construction and expression. Consider that a number of different expectations are preconsciously active at any one time and that discrepancies can occur with respect to all or any of these active processes. Such a view takes into account that even expected events that may not appear to be discrepant can be followed by emotional episodes. Consider the person who experiences the loss of a loved one who has been ill and whose death has been long anticipated. Clearly, the actual event will produce strong emotions, despite the expectation. The possible discrepancies that play into that emotion are numerous and I list just a few: The actual loss is always different from the anticipated one. After all, the other person is now truly absent, and it is unlikely that the grieving individual has rehearsed (anticipated) all the possible situations in which interactions with the living person have occurred. What has been anticipated is the actual death, not most of its consequences. Furthermore, there is usually one anticipation that is violated and that is rarely absent—the hope that the person may not die after all. And as a final example, consider the events that occur following the death of the loved one (apart from the actual loss). Friends remind one of past interactions which cannot be re-created and thus generate discrepancies; actions and plans must be envisaged that are often discrepant with previous experiences and expectations; in short, one's life changes and the changes produce emotions.[25] In general, just because one expects an event does not mean that the actual event will not bring discrepancies and surprises with it.

A similar exposition is necessary with respect to discrepancies associated with some positive events. Somehow, one may consider it peculiar that discrepancies are the source of autonomic arousal for positive as well as for negative events. One of the reasons for this discomfort is due to the common notion that discrepancy is somehow by itself a negative event, associated with frustration and other similar concepts. It is not, of course, necessarily negative, as I have already illustrated, but here I will marshal some additional illustrations. Consider the joys of young love. One has met the person of one's dreams and hopes, but reciprocity is not quite apparent. One is to meet again a few days hence, and as the object of passion appears at the designated time, joy floods the lover; ecstasy is near. What is discrepant? I would argue that the anticipation of the event is never devoid of doubts and fears: Will the loved one appear at the appointed time, is she at all interested, does he look as desirable as I once imagined? The world of romantic love is full of such ambivalences, and wherever there are ambivalences the actual event will be discrepant with some of them. There is no argument about the emotional quality, the "value" of the love. What is at stake here is the question of whether there is in fact autonomic arousal generated by interruptions and discrepancies. In contrast, consider a positive event that is fully anticipated, in all its details and nuances. For example, a cash prize has been won and the check arrives in your mail. The value is still there, but the intensity will be relatively low. I believe that an appropriate analysis of positive events will disclose the operation of many ambivalent expectancies that are more than sufficient to explain the intensity of positive emotions. Similar ambivalences operate, of course, as I have shown above, for many negative events. For every expected "good," there are thoughts of disappointments and slip-ups, and for every expected "bad" there are hopes of redemption and relief. This kind of analysis also illustrates how conditions that we usually call occasions of "uncertainty" are mixtures of positive and negative expectation of various degrees.

There is an additional difference between positive and negative emotions that is often overlooked. The result of experiencing a negative emotion is that one wishes to terminate, avoid, or leave the situation or condition that produces it; the opposite is the case for positive emotions. The additional cognitive effort required to change the condition of negative experiences preempts conscious capacity and may explain, in part, why negative emotions (in contrast to positive ones) interfere much more with constructive and productive thinking.

Evaluation: The Culture of Emotion

Difference detection is the most frequent initiator of the autonomic nervous system response that influences the intensity of emotion. The evaluative cognitions that determine the quality of the emotional experience are situated in the social context of the person experiencing them, and to a large extent in the discourse that embodies emotions socially.[26] Evaluative cognitions are, of course, "biological" to the extent that any cognitive activity is a function of the cognitive equipment that we inherit as biological products. However, the content, the evaluative part of these mental representations, is constructed by the society in which we grow up and live, and whose structure and history determine and shape our values. This combination of individual, biological, and social factors is quasi-biological, in that it provides the individual with some few prepackaged fundamental values, but there are a host of social values as well. This position argues that "emotion talk must be interpreted as *in* and *about* social life rather than as veridically referential to some internal state."[27]

A theory of the socially based source of emotional qualities also avoids the difficulties of deriving all emotional experiences from a few basic ones. Such difficulties produce Talmudic arguments about Western emotions as well as about the emotional vocabulary of non-Western cultures. The notion that all emotions are some combination of a few basic ones produces a cordon bleu school of emotional theory. If, on the other hand, one situates the quality of a particular emotion discourse in its particular culture (whether a Western subculture or a non-Western one), the task becomes one of understanding the culture rather than trying to extract recipes for complex and culturally strange emotions.[28]

Constructivist theory can also respond to the appearance of similar emotional states in different cultures. Rather than seeing such states as fear or joy as biologically basic, one can adopt an argument that has been discussed in chapter 3. Natural selection has shown that similar conditions, environments, and ecologies produce similar evolutionary solutions as, for example, in the appearance of analogically similar animal species for both marsupial and placental mammals. The same argument can apply to social evolution—that similar problems give rise to similar social solutions. Thus some emotional states may be either culturally *homologous,* having some common early human ancestry, or culturally *analogous,* having developed independently in different cultures. It is not possible at this time to decide between these two avenues for any given emotion, though it is likely that emotions that appear in some form in all cultures are homologous— that is, developed early in human culture and survived in different cultures because of the similarity of relevant conditions. One might speculate that the emotions that "basic" emotion theorists agree on (such as fear and joy) fall in the homologous category of cultural development.

As an example of different cultures constructing different emotions, consider the emotional concept called *metagu,* found in the culture of the Ifaluk, a Malayo-Polynesian group of Micronesia. Emotions in Ifaluk are described in terms of the situations that elicit them, not in terms of subjective experiences. Metagu occurs in a variety of different situations, such as when canoeing in open water, when visiting the household of an unfamiliar family, or when in the midst of a large group of people. It also describes a person's reaction to the "justifiable anger" of a person of higher rank. It is in this connection that one expects children to experience metagu when in the presence of an older person; the failure to do so is a derogation of the older person. Metagu sounds at times like the emotion we call anxiety, but it is quite different—for us anxiety is often a situationally neutral state, it can be "floating," whereas metagu is primarily an interpersonal emotion. It is something that one expects children to display, just as we expect our children to show "respect." Other emotions described by the Ifaluk that are different from the range of Western emotions are *song,* which describes an anger that is justifiable because it is in response to socially inappropriate behavior, and *maluwelu,* a somewhat passive and gentle quietude. It is difficult to empathize with strange emotional states such as metagu or song because they are defined as a function of being a member of, and having lived in, a particular culture.[29]

There are, of course, emotions that occur in all human cultures, simply because of the commonality of certain human experiences and values. These are conditioned by the very fact of being human and, in the context of discrepancy theory, of being faced with common situations and dilemmas. Among these common emotions are fears (all people are threatened by something), joys (all people face the unexpected occurrence of something desirable), and some emotions evaluatively described by common facial messages, such as disgust. On the other hand, variable emotions such as jealousy depend on a variety of different factors; even in our common Western societies emotions may be based on, or interpreted as occurring in, a variety of different situations and attitudes. The rather recent development of romantic love in the West is another example of a varying emotion that has no counterpart in some non-Western societies.

In partial illustration of my general argument, I add an example of related emotional states—the similarities between grief and guilt. The basic condition of grief (sadness) involves the loss of an object that cannot be recovered; guilt involves an act that cannot be undone. In both cases, the individual attempts to reconstitute a situation—life with the lost object or life "prior to" the unacceptable act. Both of these are discrepancies and produce SNS arousal and often strong emotional reactions. Universal emotional states are universal because of the way the human world is constructed. For example, attachment and cathexes of people and objects are a common human characteristics, and some form of grief or sadness will be found in all human societies, but the specific nature and content of the grief, guilt, etc., will be determined by the social context, including the way society defines important (love) objects and unacceptable actions.

The Contemporary Scene

For the past quarter-century there has been a veritable cloudburst of theory and research in the field of emotion. The result has been a flood of theories and little consensus among the many emotion psychologists and neuroscientists. I present here some of the most popular and representative positions on emotion as well as a discussion of two central problems in current psychologies of emotion: the question of facial expression and the reality of basic or fundamental emotions.

The general current trend has been to view specific emotions as unidimensional packages of related mechanisms. The view of "packaged" emotions is directly related to one that postulates some (usually few) basic, fundamental emotions, which elevates states such as fear and joy to a special privileged status.

Another position distinguishes between affective (emotional) and cognitive analyses. Robert Zajonc has marshalled an array of anecdotal and phenomenal evidence to argue that affective responses are unmediated[30] and respond directly to specific aspects of an event. The argument concerns preferences that are far removed from emotions as I discuss them. We live in a world of value (and affect), and the themes that determine our conscious constructions often require an affective content. This does not force an absence of other analyses and activations, such as cognitive knowledge-related ones, going on at the same time at the preconscious level. Which of these analyses will be used in conscious constructions will depend on the intentions and requirements of the moment, which happen to be "affective" in many cases. Furthermore, the assertion of "fast" and independent affective reactions is empirically testable, and the available evidence suggests that affective reactions are actually slower than "cognitive" ones.[31]

I now turn to some specific and central theoretical positions. Carroll Izard's approach is typical of the advocates of a unitary emotional complex. He says that emotion "is a complex process that has neurophysiological, motor-expressive, and phenomenological aspects."[32] Each "fundamental" emotion has its own innate program, whose neurochemical activity "produces patterned neuromuscular responses of the face and body and the feedback from these responses is transformed into conscious form."[33] At the core is an unanalyzed "innate program"—the essence of the emotion. Autonomic nervous system activity plays a part as do all the components as "part of the structure underlying the emotion process."[34] Autonomic activity is said to be patterned differentially for different emotions. When we ask what are the conditions that produce these emotion complexes, the answer lies "in the situation." There are fear situations and interest situations, grief situations, and so forth. Complex emotions are mixtures of a limited number of basic emotions. More specifically, "an internal or external event changes . . . the pattern of . . . activity in the nervous system, . . . [and that] change directs . . . [an] innately determined facial expression" that activates the emotion.[35] And once an emotion is activated, glandular, cardiovascular, and other systems are involved in its "amplification or regulation."

Nico Frijda may be the most wide-ranging and ambitious of contemporary theorists. He starts with a working definition that defines emotion as the occurrence of noninstrumental behavior, physiological changes, and evaluative experiences (or their inner determinants). In the process of trying a number of different proposals and investigating action, physiology, evaluation, and experience, Frijda arrives at a definition that is broad indeed. He describes emotion as a set of mechanisms that ensures the satisfaction of concerns, compares stimuli to preference states (and, by turning them into rewards and punishments, generates pain and pleasure), dictates appropriate action, assumes control for these actions (and thereby interrupts ongoing activity), and provides resources for these actions. The question is whether such mechanisms cover too much territory, and leave nothing in meaningful action that is not emotional. That may well be Frijda's intention, but it leaves the topic of emotion burdened with supporting practically all of psychology.[36]

Ortony, Clore, and Collins are straightforward. They define emotions as "valenced reactions to events, agents, or objects, with their particular nature being determined by the way in which the eliciting situation is construed."[37] Such a definition is subject to James's critique; it is abstracted from the "bodily felt" emotions. But as a definition of "affect" (the cognitive part of the emotions), it is the most consistent and consequential approach.

Keith Oatley and Johnson-Laird's theory of emotion is the only one that specifically claims direct ties to the computational cognitive science enterprise. Their approach is more complex and elaborate than most, and it introduces new terminologies (though often for old concepts). They propose a system of modular processors in the human information-processing system with emotion modes that are nonpropositional communications setting the system for appropriate action, including switching appropriate modules on and off. These nonpropositional signals can function without higher-level cognitive evaluations and without conscious intervention. There are five basic emotion modes (in keeping with other basic emotion models). Complex emotions are not mixtures of the basic ones, but cognitive elaborations of them. In addition, the emotion modes coordinate the modular nervous system, and the cognitive system "adopts an emotion mode at a significant juncture of a plan."[38] These junctures are the equivalent of particular cognitive structures specific to the five basic emotions and as cognitive structures are not much different from the kind of structures envisaged by Ortony et al. For example, for anger the "juncture" is "active plan frustrated," and a transition occurs to a state of "try harder, and/or aggress."

Richard Lazarus and his co-worker Susan Folkman define emotion as organized reactions that consist of cognitive appraisals, action impulses, and patterned somatic reactions.[39] Emotions are seen as the result of continuous appraisals and monitoring of the person's well-being. The result is a fluid change of emotional states indexed by cognitive, behavioral, and physiological symptoms. Central to Lazarus's position is the notion of cognitive appraisal, which is an integral part of the emotional state, and it leads to actions that cope with the situation. Coping is an important concept in this position, and it can be centered on problems faced or on emotions experienced. Primary appraisal asks what is at stake in a situation and defines the quality and intensity of emotion, whereas secondary appraisal asks questions about how to cope with a stressful situation and about the response of the environment to such reactions.

Do Facial Expressions Express Emotion?

Apart from observations of daily experience, the linking of emotions and facial expression has its origins in Darwin's *Expression of Emotion in Man and Animals.* Unfortunately, the linking of Darwin's theories and facial expression has left the impression that Darwin considered these facial displays as having some specific adaptive survival value. In fact, the major thrust of Darwin's argument is that the majority of these displays are vestigial or accidental or, at best, preadaptive. In fact, Darwin specifically argued against the notion that "certain muscles have been given to man solely that he may reveal to other men his feelings."[40] Fridlund has explored Darwin's 1872 motive and message and notes that his antiadaptationist view of facial displays also prevented Darwin from viewing these displays as primarily communicative.[41]

The contemporary intense interest in facial expression started primarily with the work of Sylvan Tomkins, who placed facial expressions at the center of his theory of emotion with eight basic emotions forming the core of emotional experience.[42] The work of both Paul Ekman[43] and Izard derives from Tomkins's initial exposition.

The notion that facial displays express some underlying mental state forms a central part of many arguments about the nature of emotion. Apart from the fact that it needs to be made clear how the outward expression of inner states is adaptive—that is, how it could contribute to reproductive fitness—important arguments can and have been made that facial displays are best seen (particularly in the tradition of behavioral ecology) as communicative devices, independent of emotional states.[44] Facial displays can be interpreted as remnants of preverbal communicative devices and as displays of values (indicating what is good or bad, useful or useless, etc.). That position is more or less identical with the original work of Fridlund,[45] who notes, inter alia, that facial displays are consonant with current evolutionary views of signaling and that even displays previously considered involuntary are in fact social and communicative. Fridlund has shown how a social interpretation of these displays best fits with existing knowledge about the function of displays in emotion and their presumed universality. He has elaborated a scenario of the evolutionary origin and utility of facial displays in which these displays function to communicate intentions and situational evaluations in the absence of verbal devices. The work of Janet Bavelas and her colleagues has also shown the importance of communicative facial and other bodily displays. Their conclusion, in part, is that the "communicative situation determines the visible behavior."[46]

In the construction of emotions, facial displays are important contributors to the evaluative cognitions and appraisals of the current scene, similar to verbal, imaginal, or unconscious evaluative representations. Facial displays occur in many situations where emotions are inferred or asserted. However, they also occur in many situations that one would not call emotional at all; facial and body language frequently provides important social communications. In fact, the original position of some inexorable link between face and emotion has been softened in recent years, as when Ekman noted that facial expressions may serve a variety of different purposes, including the transmission of information.[47] Furthermore, if emotions are conceptualized as a concatenation of evaluative cognitions and sympathetic arousal, then seeing facial displays as displays of values can make them often (but not always or inevitably) part of the emotional complex. I leave open for the time being the evolutionary history of these displays and their relationship to the apparently similar displays of nonhuman animals that so fascinated Darwin.

The Question of Basic Emotions

An important part of the argument for the impenetrability of emotions is the postulation of basic or fundamental emotions. Constructivist approaches have usually rejected such a view. The lead for such a rejection was taken by James, and it has received important support from recent detailed expositions.[48] Rejection of basic emotion does not prevent one from looking for basic elements that constitute emotions, except that such basic elements are not in themselves emotions. Even though many tales have been spun about the evolution and origins of the separate emotions, there is little agreement or consistency to be found.[49] One of the difficulties that faces speculators about the origins of discrete emotions is that they have not yet agreed on what the discrete fundamental emotions are.[50] Thus, Ortony and Turner note that the number of basic emotions can vary from 2 to 18 depending on which theorist you read. If there is an evolutionary basis to the primary emotions, should they not be more obvious?

The emotions that one finds in most lists of basic emotions are fear/anxiety, happiness/joy, and anger. Again, the list is heavily weighted toward the negative emotions. Two "emotions" sometimes included are interest and surprise as distinct and separate emotions. To call surprise an emotion depends on one's interpretation of the common usage of "emotion." And since many different emotions such as fear, happiness, etc., may involve some degree of surprise, how does one deal with surprise as a separate emotion? On the other hand, surprise is an excellent example of the reaction to discrepancies. To insist that interest is an emotion is a more extreme position. There seems to be little basis in experience or theory to consider the expression of interest indicative of an emotional state; to call interest an emotion moves such a position far from the general understanding. On the other hand, it is equally puzzling that the emotion of "love" (much less "lust") is never found among the basic emotions. Is it because no distinct facial "expression" can be found for love? The best current summary about the role of facial expressions is that they are more complex than frequently believed.[51]

Some Speculations about the Evolution of Emotion

Discussions about the evolution of the emotions are often preceded by an argument that our emotions evolved to fit an environment that was wild, dangerous, and uncivilized,[52] or that large portions of our animal heritage "have become inappropriate to civilized life."[53] The focus is usually on the "violent" emotions, such as pain, fear, and rage, which are "out of place." But one tends to forget other violent emotions, such as the positive passions of love (whether of sexual partners or of one's children) and lust, ecstasy, and joy.[54] Are these too "out of place"?

What is often forgotten is that the environment to which we originally "adapted" by developing these emotions was designed by God, nature, chance, or whatever. However, the very environment for which we are said to be unfit emotionally was designed by the very organism that does not fit it. I wonder whether an argument that says that it is unlikely that cultural evolution would design an environment precisely unfit for the designer is not just as viable as the argument that says we are in fact unfit for the present environment. And in what way are we unfit?

The notion that specific emotions evolved somehow separately for different reasons requires repeated appeals to their adaptive origins. Whereas arguments have been advanced for the adaptive uses of emotions in general, frequently specific evolutionary stories have been advanced for different emotions. However, appeals to adaptation (the Panglossian argument) often ignore the fact that current advantages (or disadvantages) may have relatively unrelated evolutionary origins. More important, current utility does not permit inferences about evolutionary history. And since it is not clear how one is to use the term "emotion," it seems somewhat futile to describe how it may have been selected and how it has evolved. If, on the other hand, one defines emotions as evaluative and autonomic conjunctions, then the role of the autonomic nervous system plays an important role.

One of the difficulties in exploring the evolution of human thought and action or even in guessing at evolutionary developments that have temporarily produced current humankind is our fascination—clouded by ethnocentric and sociocentric concerns—with the existing phenotypes, and even worse the phenotypes that we happen to find in our immediate environment. To find, as is often done, that the phenotype seems to be well adapted to current conditions does not, of course, say anything about the genotype that contributes (in some undetermined part) to the observed phenotype.

The overdetermined nature of much of evolutionary development is well represented in the distinction between particular behavior patterns and their proximate causes—the underlying factors that cooperate in producing that behavior. I have illustrated that argument in chapter 3. Such an insistence on underlying mechanisms also informs constructivist approaches and calls for an integration of the study of proximate underlying mechanisms and their ontogenesis with the functional or evolutionary approach.[55] Constructivist approaches to emotion similarly look for an integration of underlying mechanisms (e.g., discrepancy and evaluation) with the broader functions of emotions. Basic processes such as discrepancy and evaluation can produce a variety of emotional states, and a search for specific selective processes for specific emotional states is probably ill informed. Unfortunately, speculators about emotion in particular have too often been concerned with finding a unique evolutionary basis for the functions of complex phenotypes.

In trying to understand the evolution of the physical and psychological bases of subjective emotions, I start with the generally agreed viewpoint that the autonomic nervous system developed for primarily economic reasons, first by the evolution of (parasympathetic) energy storage functions, followed by the (sympathetic) energy expenditure functions.[56] At some later time (or even in parallel) there developed the discrepancy or interruption effect on sympathetic arousal, and interruptions of action became sufficient occasions for SNS activity. The evolution of the SNS and its functions occurred quite early in the development of mammals. Much later, prior to the emergence of language, but as part of the development of modern humans, we place the evolution of the cognitive apparatus that informs evaluations. The initial and early steps in that evolution were the development of specific response tendencies and action syndromes. These include some vocal patterns, motor patterns such as reaching and avoiding/flinching, and sexual behavior. The next evolutionary step was probably the emergence of mental representations of these actions, which are the "affective" cognitive evaluation of actions and situations. These evaluations also include the feedback of automatic involuntary actions, where self-perceptions, when we see ourselves approaching or avoiding objects and events, become the basis of evaluative reactions.[57] The last step, in the nonverbal environment, was probably the development of a bodily and facial language, not to "express emotions" but rather to communicate evaluations—warnings, encouragements, etc. I assume that all these developments took place before the emergence of verbal language and its particularly powerful ability to encode and communicate evaluations and values. The combination of these initially separate and independent tendencies and characteristics eventually combined to produce the mechanisms and processes that construct the emotions. The SNS provided the intensity, and various actions and communicative behaviors added the quality of the emotional concatenations. The unity of consciousness, the holistic nature of conscious experience, combines these separate developments into the experience of emotion.

The "evolution" of emotions then becomes not a separate unique evolutionary event, but the fortunate (or unfortunate) outcome of the independent evolution of its constituents, eventually combined in a conscious animal into unitary phenomenal experiences. Emotions need not be seen as adaptive or selected for some contemporary reason or another; we have been bequeathed these outcomes as a result of a variety of evolutionary developments, just as we have been bequeathed a writing instrument when the hand developed.

SEVEN

Values in Human Thought and Action

Origins and Functions

Beauty in things exists in the mind which contemplates them.
David Hume

ALL HUMAN SOCIETIES, and all members of our species, have values. We all have preferences, we all make judgments that something is good or bad (or neutral), we all have tastes about food, clothes, weather, and so forth. In addition, as I have discussed in the previous chapter, many of the affective and evaluative states that are frequently subsumed under the rubric of emotion are primarily statements of value. I turn now to an analysis of these values.

Apart from mainly descriptive studies, or philosophical disquisitions, the academic disciplines have been comparatively quiet about the origin and development of values. Psychology is particularly guilty of pretending that value is not a proper subject of study, either as a problem of its psychological representation, or as reflected in psychology itself. This avoidance has been incomprehensible to some and comfortable to others, and one psychologist noted the lack of attention to value by predicting that "[i]t is quite possible that future generations will look back upon this period in utter perplexity."[1] This expectation of future perplexity is a quote from a paper written over 50 years ago! Wolfgang Köhler believed that psychologists in the first half of the century refused to consider problems of value because, as Köhler quotes his colleagues, "scientific psychology . . . deals with strictly neutral facts, just as does physics."[2] Philosophers of science and of society have since agreed that there are no "neutral" facts, but psychologists have still avoided the problem. The dissection of psychology and psychologists that would be necessary to understand the social and scientific matrix that created this phobia could constitute a lengthy chapter of its own.[3]

I do not intend to address the issue of innate basic value orientations—whether human beings are a priori good or bad, aggressive or cooperative. Thus, I do not speculate about Panglossian perfection or Augustinian evil. I believe that we now have enough sophisticated evidence both from anthropology and primate studies[4] to realize that human and simian diversity place relatively weak constraints on the range of possible human values. Nor do I intend to analyze in detail the expression of the more obvious "built-in" behaviors that are usually interpreted as indications of value. Human beings, from birth on, like sweet tastes and retreat from looming objects. They startle at loud noises, cling to supporting objects, and avoid loss of support. These, and several other, avoidances and withdrawals form the basis for many human values, but in themselves they account for a very small proportion of the values that govern our social lives. Liking candy, shrinking from heights, and avoiding thunder inform little of our everyday lives.

Most of my exposition is weighted toward what might be called "simple" values. I am interested in the origins of people's calling an event pleasant or noxious, an experience joyous or frightening, and how we come to have preferences for certain foods, music, paintings, etc. In that sense I tend toward an explication of tastes rather than values in the sense of moral, rational values (and their justification). There are complex social values and institutions that are beyond the reach of this discussion, such as values of justice, honesty, and courage. In many cases, the discussion to follow touches on these cases; in other cases, these values speak primarily to behavior and action that are approved by a particular society or subsection thereof, and some are discussed in chapter 10. Simple values, as I use the term, are values that are frequently not debatable—one either likes or does not like clothes, pictures, people, foods. Complex values, on the other hand, are debatable and sometimes require justification. Why is it good to be humane or cooperative, to believe in democratic values, to assert one's opposition to capital punishment? Sometimes these values, too, are held uncritically and automatically, but they also generally fall into the category of moral values. The latter have been the philosophers' province, those philosophers who usually wish to construct or discover systems of moral judgments that are *rationally* defensible—that is, values for which reasons and justifications are available and accessible. No intent to explicate such complex values should be read into the following sections.

I also try to avoid defining values in terms of other psychological phenomena. I want to get at fundamental psychological processes that are intrinsic to the development and construction of values. In contrast, some philosophers, as in the utilitarian tradition, have tended to approach values in terms of wishes, strivings, or interests, trying, for example, to define values in terms of valuable objects of interest or in terms of the interest taken in them.[5]

Most of my discussion is concerned with the structural basis for the persistence and change of values, which I show to be a function of the structure of the schemas that inform our view of the world. I contrast the way these schemas define such aspects as familiarity and novelty with "social" values that determine the contents of the schemas. I choose the term "social" to indicate that most of our values arise directly out of the character and content of the social order. In juxtaposing primarily structural mechanisms and mainly contextual social ones, I do not imply that structural, schematic mechanisms are independent of the social matrix. Obviously, the social context determines what schemas can be constructed. However, neither for the structural argument, nor the social one, do I address the mechanisms whereby social values are acquired. I avoid this topic primarily out of ignorance. I do not believe that there is just a single learning mechanism, or even that the basic "laws of learning" are generally known. We understand some of the conditions under which new actions and beliefs are acquired, but we know too little about the varieties of learning mechanisms to make even an approximate statement about the acquisition of values, beliefs, or attitudes. I do, however, address the conditions in which these acquisition processes operate—the social structures that make one or the other value more or less likely to be acquired.

I remind the reader of an important point, especially relevant to the problem of values: Social scientists have been prone to propose and look for single dimensions of causation—single causes and unique deterministic chains. I too find myself at times a victim of this particular social disease. But values are not unidimensional. Psychologists in particular have advocated monolithic mechanisms for such complex phenomena as learning, perception, and intelligence. In part this has been a function of the psychologists' successful use of the experimental method, which has often been most useful in isolating the effect of single variables and processes.[6] However, the human organism not only responds complexly to the world, but it is also a redundant fail-safe system—there are a variety of processes and mechanisms that address a particular problem of adaptation and environmental demand. There are, for example, many learning mechanisms and there are a variety of ways of generating values and the experience of value. In the world of daily experience and social intercourse, values are multidimensional and overdetermined. Not only are there different sources of value, but the expression in action and thought of some value orientation is usually derived from more than one source.

I start with a short discussion of the definition of value. Following that, I briefly recapitulate my discussion of emotional experience, because the problem of value is at the heart of human emotional experience. I do not assert the converse, that emotion is at the heart of the problem of value. Values, even simple preferences, do not arise out of emotion; instead, they contribute to emotional experiences. One might consider emotions to be a common language metaphor for the intense expression of values.

After developing the theme of the role of values in the experience of emotion, I concentrate on what I believe to be a central condition that gives rise to value attributions and experiences: the emergence of stable schemas that create a conservative view of the world on the one hand, and contradictions among these schemata that produce novelty and changes in values on the other hand. The discussion of discrepancies and values harks back to the difference detection mechanism discussed earlier, while the exposition of familiarity is concerned with the other pole of experience—the experience of an unchanging world. I am concerned with structure, just because it is the schematic structure as such that gives rise to a sense of familiarity or acceptance and also determines the sign or direction of consequent values when schemas are interrupted or inconsistent with current evidence. I then continue with some thoughts about the social contexts that define and generate values. For analytic purposes, I initially discuss the three sources of values—innate, structural, and social—as independent phenomena. I include in my conclusion the assertion of their interdependence.

What Is a Value?

Despite quarrels about definitions that have been one of the hallmarks of the social science enterprise, one can find general agreement that what is needed for a psychological approach to value is some representation that shapes our likes, dislikes, preferences, prejudices, and social attitudes, that informs (but does not constitute) our moral judgments and, in general, that makes it possible for us to say what is good and what is bad. Generally speaking, we exercise simple values unconsciously. We know what we like. Choices and preferences come unbidden and usually without deliberation. We know what people we like, what foods and works of art we prefer, and we know these things automatically, without reflection. And we "exercise" many of our social values of competition and cooperation, humanism and racism, altruism and self-aggrandizement, patriotism and chauvinism in a similar fashion. I recognize the fact that many of our complex moral values are "rational"—that is, they are debatable and often require or bring forth justifications. On the other hand, we often fail to inquire into the state of our values in everyday life and, when challenged, often rely on social norms or rationalizations. On one hand, our response may be: "I just do"; on the other hand, we may have reasoned justifications that may be no more than appeals to socially acceptable formulas, given without further examination. Some of this latter behavior is best described as "false consciousness"; people believe that they know what values they are exercising, and insist on (moral, ethical) bases for choices that are often no more than rationales generated by institutions and groups through their dominance of the social means of communication. In general, society and the norms produced by it not only generate rationalizations but also constrain the possible values and system of values to which anyone can appeal as justifications for actions or ideas.

For a definition of value, one might adopt that proposed by Milton Rokeach, who has contributed more than most to the psychological study of value. Rokeach defines values as enduring prescriptive and proscriptive beliefs in the preferability of modes of actions and of end-states.[7] A value is a standard that guides actions in much of the individual's everyday life, and values are hierarchically organized along a continuum of importance.[8] As to how these values are specifically acquired, Rokeach is no more helpful than others. He notes that values are "the result of all the cultural, institutional, and personal forces that act upon a person throughout his lifetime."[9] How these "forces" produce personal values is left open.

Köhler is the one psychologist who has made a heroic effort to place value squarely within psychological science. His William James lectures in 1934–35 placed value within the notion of requiredness—values are demanded by the structure of both the phenomenal and the physical world.[10] Value is an instance of the recognition of requiredness that is determined by the *Gestalten*—the inherent patterns of our physical, neural, and phenomenal worlds. Facts do not just happen or exist, but they extend in specific contexts toward other facts with a quality of acceptance and requiredness.[11] The world consists of segregated contexts that have physical and valuative properties that are displayed as contexts and systems; that is, properties such as values are characteristics of structured contexts. Furthermore, any part of a context has properties (such as values) that are determined by the position of that fragment as part of the larger context or system. These contextual systems change and develop historically, but such change is not subjective—for example, imposed by the individual's phenomenology. Rather, historically changed systems (as, for example, the acceptance of minor chords in music) are as objectively real as the preference for sweet over bitter substances.

There are certain appealing qualities to a view of the world as determined by objective structures, but Köhler wants to avoid subjectivity to the point of caricature. Consider the following quote from a later paper:

> [Irresistible womanly charm] is a value attribute on which women have a monopoly. It would be absurd to maintain that when the intensely male interests . . . impinge upon the neutral appearance of women, female charm develops in these objects as an illusory projection of the males' conations."[12]

I believe many psychologists today would opt for "absurdity." This does not deny that there are "objective" characteristics of female (and male) sex objects, qua sex objects, nor that there are historical and social standards of beauty, but pulchritude (and its various social and cultural variations) is surely "in the eye of the beholder."

Emotion and Value

As I have noted, my concern with spelling out a psychology of value derives directly from my discussion of emotion in the previous chapter.[13] To recapitulate, my approach addresses the subjective experience of emotion. It is not primarily concerned with emotional behavior, which may or may not be accompanied or followed by positive or negative emotional experience. I have focused on two dimensions of the many available from analyses of common language "emotions": the notion that emotions express some aspect of value and the assertion that emotions are "hot"—they imply a gut reaction, a visceral response. The cognition of values, what is good or bad, provides the quality of the emotional experience, and the visceral reaction generates its quantitative aspect.

Sources of Values

I start with the three classes of values described briefly in the previous chapter: those arising out of actions and behaviors, socially acquired values, and structurally determined values. The "action" and "social" categories determine the contents of value structures, while the structural factors are primarily responsible for the persistence as well as the change of values.

In the case of actions, individuals often interpret their automatic actions and behaviors in value terms. These include innate approach and withdrawal tendencies such as the avoidance of looming objects, the avoidance of pain (and pain-producing objects), and an approach to sweet substances. It is the secondary effect of these tendencies, such as our observations of our own approaches and withdrawals, that is one of the conditions that produces judgments of positive and negative values. The observations of our own actions may inform the value that is generated in a situation.[14] The experience of value arises out of our perception of our approaches and withdrawals; if one sees one's hand recoiling in pain from a hot surface, one may learn to call scorching objects "bad." However, it is equally possible that approach and withdrawal themselves are the expressions of value.[15] These actions are frequently innately determined, but in the adult, a large variety of approach/avoidance tendencies have been acquired and form the basis of this class of values.

A second source of affective values are cultural, social, and idiosyncratic predications—the social values. The social context determines to a large extent how various objects and events are to be represented and how they are valued. Events and objects, whether actually encountered or not, are predicated to have certain values as a result of social or personal learning experiences and contexts. Food aversions (such as to spinach or liver) are frequently acquired without any contact with the actual substances, as are likes and dislikes of people and groups of people. Culturally acquired aesthetic judgments of beauty, whether of people, landscapes, or paintings, may have similar origins.

The third source of value in the construction of emotion is structural. There are three different structural contributions to the emergence of values: patterning, consistency, and discrepancy. Patterning addresses the cognitive structure of objects and events and depends on the relations among features and attributes rather than on the presence or absence of certain features. Scenery, paintings, and people can be seen as beautiful or ugly as a function of a particular structural concatenation, a particular pattern of their constituent features. Value may arise out of differentiating patterns, in contrast to the mere identification of objects or events. The valence of a specific patterned object or event is determined in a large part by our experience with them and our analyses of their constituent features. Except for some aesthetic patterns, it is the social context that defines which pattern will be judged as positive or negative. Thus, it is the patterning of sounds that describes acceptable music for a particular culture, and it is the pattern of the features of the human body that determines local standards of beauty. Feature patterning is a representation of value, whereas the two aspects of structure to be discussed next—consistency and discrepancy—are more important for the generation of values.

Familiarity, Discrepancies, and Contradictions

One of the factors that influences judgments arising out of structural consider-ation is the frequency of events and encounters. The more frequently an object or event has been encountered, the more consistent the representational schema. If an experience fits an existing schema, it is, *ceteris paribus,* an acceptable or ex-pected event.[16] In most cases it is the preferred state of the world. We generally prefer the known to the unknown. If we see an object, whether person, animal, or structure, that conforms to our acquired views of what such an object should look like, we tend to like it.

The available evidence supports the hypothesis that sheer exposure and mere frequency of encounters generate familiarity, acceptance, and schematic "fit." However, it is possible that the sense of familiarity and acceptability arises out of the fact that repeated exposures make an event comprehensible and understand-able. Such comprehension arises in part out of consistency with expectations. According to such a view, it is the acquisition of knowledge about a domain of experience as well as its fit with a model of the world and one's environs that generate comprehension and acceptance.

There is another aspect of familiarity that may well have evolutionary bases. On average, unfamiliar persons, animals, situations, and objects may be danger-ous and thus to be avoided, and such an avoidance could well be part of our evo-lutionary heritage for its obvious selective advantage. Such a trait would under-line and reinforce the preference for the familiar.

Frequent experiences with negatively valued events appear to present a para-dox for this approach. Do we eventually like what we initially disliked? The ex-perimental evidence is contradictory: Repeated exposure to initially negatively valued events may increase or decrease the negative evaluation.[17] Everyday ex-periences with novel and unfamiliar music, food, arts, and people suggest that extended experiences may change initially negative reactions to positive ones. More important, the negative value of a particular event—its noxious aspects—is often maintained together with the familiarity or acceptance value generated by frequent exposures. The aversive nature of living in a fascist state or endur-ing a malicious neighbor coexists eventually with an acceptance of the status quo. Such a situation generates one of the more pervasive ambivalences of human life—the apparent acceptance of a basically unpleasant way of life. Stories of long-time prisoners who are reluctant to leave jail after many years illustrate one of the more extreme examples. Values are not unidimensional, and many different sources of value generate a multifaceted way of perceiving the world.

Schema theory generates a psychological picture of humans that makes com-prehensible their conservatism as well as their acceptance and generation of change. Order arises in part out of repeated experiences with our personal world, giving rise to a conservative organism who prefers the known to the novel, the old to the new.

On the other hand, the world also changes, and difference detection, the re-action to discrepancies discussed in chapter 6, comes into play. The world is not constant—it presents us with changed conditions, reordered social relations, power positions, and dependencies. Change is often inherent in the structure of a society that generates inconsistent expectations and contrasting values and requirements. Such social inconsistencies are often referred to as social contradictions. Contradictions imply change and emotional involvement. Discrepancies and disorders are an inherent aspect of our world; we attempt to impose structure and order on the disorder, and the attempted resolutions of contradictions in our lives lead to new discrepancies.

Schemas change in the course of one's interactions with the world, and do so dramatically in early life. For the adult, schemas become powerful organizers of expectations, beliefs, opinions, and actions. Eventually, little change occurs in our perception of the social environment and even less for the physical environment. Normally, the world changes relatively slowly and schemas easily assimilate changes in our world of communication, production, transportation, etc. Television did not suddenly replace radio, and automobiles only slowly displaced horses. The social world is somewhat more changeable; we encounter new people, make new friends, construct new conditions of work and play. Accommodation and assimilation describe the major agents of schematic change.

Our species operates apparently on a principle of least effort when it comes to schematic change. Current schemas—our current view of the world and ourselves—are changed only with effort, and, whenever possible, we stay with what we know and with past actions and beliefs. The degree to which an existing schematic view of the world will change depends on the degree of consistency and integration of the schema on the one hand and the challenge that new information provides on the other. Accommodation requires cognitive effort, and in the absence of some motive for change, we stay with current conceptions and perceptions. The motives for change may be the availability of alternative views of the world that promise better conditions of existence or intolerable current conditions that require change. Habitual ways of knowing and perceiving are typically preferred to novel constructions that require mental (and often physical) effort. Assimilation is the preferred mode of dealing with the world and, I argue, is in fact one of the basic values that informs human thought and action. What is known and expected is, *ceteris paribus,* preferred to what is unknown and unexpected. The more experiences we have had with a particular event (be it a person, a home, a career, or a food), the more likely it is that we will find it acceptable. An extensive experimental literature has shown that repeated encounters (development of stable schemas) increase the acceptability of and preference for the repeated event. As I have noted, such encounters also affect the experience of initially disliked objects. A not surprising demonstration of the parallel between knowing and liking was demonstrated in a study in which undergraduate students made judgments of whether they knew—had seen before—and whether they liked (or disliked) slides of paintings. The slides were selected from the Renaissance, nineteenth and twentiety century (mostly impressionists), or modern abstract periods. Knowledge decreased precipitously from 66 percent to 24 percent as a function of the chronology, and liking also decreased for the three periods.[18]

Another way of approaching the same issue is to consider the basic schema of an event or object, together with its default values, as the prototype of that kind of event.[19] To the extent that an event "fits" that prototype, it is considered typical of that class of events and also preferred or liked.[20] Similarly, a study of unfamiliar examples of insects, birdsongs, seashells, and paintings found that judgments of typicality were highly correlated with preference judgments.[21] Of course, correlation does not necessarily imply a causal relation between typicality and liking — it could be that things were judged as more typical because they were liked. In any case, these findings support the idea that we tend to like things we perceive as typical of their kind.

I have asserted that if an experience fits an existing schema, it is positively valued. There are, of course, many situations where we do not seem to like or prefer the familiar. Upon analysis, it is usually the case that other sources of value override the mere structural fit. But when nothing but structural comparisons predominate in a situation, we will find the preference for the familiar. One possible set of conditions for boredom exemplifies the intrusion of other values on the preference for the familiar. It is often the case that boredom is experienced when a particular event, often somewhat repetitive, produces a negative reaction because the object or event fails to conform to some standard or expectation. Boring lecturers repetitively tell us what we already know or what they have already said before; boring movies fail to conform to the expectation of being entertaining; boring clothes fail to respond to a need to be trendy.

I have struggled with the problem of defining—that is, of finding just the right expression for—the sense of preference or liking that arises out of the familiar. We really do not have the right words for it, but it seems to be related to a sense of comfort, nostalgia, a sense of belonging, and similar "sentiments." It is a sense of avoiding cognitive effort—the cognitive structures don't need to be changed—and we passively accept their familiar contents. Consider another more unusual example of the familiar being the accepted and acceptable. Abusive parents have sometimes been the victims of parental abuse during their childhood. Abuse and the violence of the parent–child interaction are the regnant schema; they are the familiar aspects of the family situation. Parental love and security are intimately tied to the occurrence of abuse—being a victim is part of the implicit and accepted definition of being a child. Similarly, masochistic experiences and needs also represent a sought out concatenation of pain with love and acceptance.[22]

So much for the conservative, status quo maintaining aspect of human nature. Our social world, in particular, is often a world of discrepancies. What we have experienced and expect is contradicted by new social conditions, by changes in the social order. The contradictions introduced by the economic, social, and interpersonal world are the very discrepancies that generate arousal.

If we accept that discrepancies and contradictions abound in our world, then we need to consider how these discrepancies are resolved. The dominant model in psychology has been one based on simple tension reduction and the achievement of orderliness and quietude, as Freud suggested. Another approach has been to borrow Cannon's model of the homeostatic resolution of imbalance and to apply it to psychological states. Rudolf Arnheim, however, tackled the problem of order and disorder from the point of view of Gestalt psychology, rejecting both simple tension reduction and homeostatic models of disorder resolution and opting for the achievement of a new structure. He noted that, in contrast to mere orderliness, the imposition of order in the face of disorder (and tension) requires active processes; the individual must effortfully find and impose a new ordering on percepts and concepts. Disorder "is not the absence of all order but rather the clash of uncoordinated orders."[23] In terms of the present discussion, disorder implies contradictions, which suggests that discrepancy and contradictions require accommodation—that is, the establishment of new schematic structures.

Within the context of emotional reactions, one can assign possible resolutions following different degrees of discrepancy or discontinuity. Consider a continuum starting at severe incongruity where assimilation (into an existing schema) is not possible and accommodation (change of schema) is necessary. Such a change can be successful when the new schema "fits" the new experience. If successful, the very fact of successful accommodation may produce a positive value and, in addition, tension or arousal is presumably ameliorated and the value will be positive. On the other hand, the incongruity may produce a negative tone, and in some cases intensely so—for example, if the accommodation requires the forceful removal of a disruptive event (as in the case of aggression). Conversely, such events as unexpected praise or recovering a lost person or object would lead to a positive evaluation. In the case of unsuccessful accommodation, I assume that the sign of the subjective value will be negative. What the specific negative value will be depends on the context of the event and past experience with the event itself. If, however, a previously established alternate way of handling the situation is found (e.g., as in denial), arousal is attenuated, and the value of the event is likely to be positive. In the cases of less severe discrepancies, where assimilation is possible, I assume that all outcomes involve no further mental effort, and the values tend to be positive. The positive value arises at least in part out of the fact of successful assimilation—the very act of assimilating an experience may well be positively valued. I have also suggested that a parallel of such assimilation—that is, successfully completing an intended action or thought sequence—has positive value, resulting in what might be called the "joy of completion."[24]

The preceding discussion suggests that accommodation involves the generation of a new structure, a new ordering of experience. But, we should note that the imposition of structure (and order) on an event or experience does not necessarily produce a state of quiescence. The newly imposed or discovered order will bring its own discrepancies and contradictions. For example, Arnheim notes that "[o]rder and complexity are antagonistic, in that order tends to reduce complexity while complexity tends to reduce order."[25] Complexity requires coping with a variety of different attributes and relations among them, and as a result, structures may change and reduce the ability to cope with the event. New structures may also bring with them values of their own. The affective state following a state of discrepancy depends on the evaluation of the new, changed state. Rarely does the perceiver discover the optimal order with a structure and complexity that can be accepted unchanged and unchallenged. Depending on the particular context and the person's preparation for the new situation, a variety of different reactions is possible. If past experience predisposes one to deal effectively with new complexities, then positive values are generated. If one feels helpless in the face of changing conditions, then values will be negative. In some cases, the complex structure may be preferred just because it invites new explorations, new tension, and new contradictions to be resolved in ever new perceptions of the world.

The Challenge of Novelty

Novelty is not an all-or-none concept; an event may be novel because some well-known evidence is encountered that has not been encountered in a particular context before, or it may be novel because it is entirely "new," and a number of variations of these themes all deserve the label of novelty. Frequently the novel produces negative affect—it does not fit and there is no way in which we can accommodate to its demands. New styles of music and painting, new ways of structuring our social world, and new modes of organization are usually greeted with negative evaluations. When is the novel positively valued—as it often is?

One possible explanation for the search for novelty lies in the positive evaluative state when a novel event or experience can be assimilated into existing schematic structures. The positive affect generated by smooth assimilation would encourage the search for novel, but not extremely discrepant, situations.[26] Another candidate for perceiving the novel as positive is some other evaluative structure that interacts with the novel event. Such structures may be social, as when a novel aspect of a difficult problem is seen as a possible solution; or it may be individual, as when people seek the novel and find positive value in problem solving, whether cognitive or aesthetic.[27] The creative artist and scientist are both part of a cultural tradition that values the novel construction and that seeks out novelty. On the other hand, there are social and cultural conditions in which the novel is avoided and considered inappropriate (e.g., in authoritarian societies).

There may also be phyletic differences, since most animals react negatively to novelty; i.e., they are unable to accommodate to the new situation. Young humans and other animals, on the other hand, apparently seek out novelty as a major mechanism in understanding and mastering their world. Whether or not we are an information-seeking species as such is still open to argument. The basic preference for the old emerges in children after schemas about the world have been well established as, for example, in stranger anxiety. New information is sought whenever the physical and social world changes and requires new adjustments. In the modern industrial world we apparently have become novelty junkies, but in the light of very stable societies (as in some parts of Micronesia), it seems premature to generalize our own experience to an assertion about human nature.

The novel and particularly the creative event produces discrepancies and discontinuities. New ways of doing and seeing are by definition of this order, but the creative act often is seen as destructive and alien. This is usually the case in social action, but it is most easily demonstrated in the arts, where the new way of seeing or hearing creates strong dissonances in the viewer and hearer. Beethoven's music provides us with many examples. A contemporary critic of the *Eroica* symphony noted that one left the concert hall "crushed by the mass of unconnected and overloaded ideas and a continuing tumult by all the instruments."[28] The effect of novel creations on an audience is contrasted with the peculiar function of creative individuals in producing discrepancies and new values. Among other characteristics, they need to accept the necessary destructiveness of their actions.[29]

There are reasons other than structural change why discrepancy and contradiction foster value-laden reactions. I have discussed in chapter 6 the possibility that the sympathetic nervous system functions in part to alert the organism to novel situations.

The import of the preceding is that discrepancies and contradictions are the occasion for the development or imposition of new values, that changes in values are most likely to occur when contradictions have to be faced. This is to be contrasted with the conservatism of holding to the familiarity and comfort of well-established, habitual schemas. Which of these two roads will dominate in a situation depends on the general system of schemas that characterize the individual's value system. The outcome of the dialectic between conservatism and change is in turn determined by historical and social aspects of the society and the personal history of the individual within that society.

The Arts and Discrepancies

The most intense interest in the role of discrepancy and discontinuity and their effect on value has developed in the arts, both in visual arts and in music. I have already referred to the work of the noted psychologist of art, Arnheim. Another commentator on the visual arts, art critic John Berger, suggested a pervasive disjunction or discrepancy in art:

> Before any impressive painting one discovers the same enigma. The continuity of space (the logic of the whereabouts) will somewhere on the canvas be broken and replaced by a haunting discontinuity. This is true of a Caravaggio or a Rubens as of a Juan Gris or a Beckmann. The painted images are always held within a broken space. Each historical period has its particular system of breaking. Tintoretto typically breaks between foreground and background. Cezanne miraculously exchanges the far for the near. Yet always one is forced to ask: What I'm being shown, what I'm being made to believe in, is exactly where? Perspective cannot provide the answer. The question is both material and symbolic, for every painted image of something is also about the absence of the real thing. All painting is about the presence of absence. This is why man paints. The broken pictorial space confesses the art's wishfulness.[30]

In the analysis of musical aesthetics, Leonard Meyer is the most emphatic modern writer in tying affect directly to the interruption or inhibition of expectations. Meyer's discussion is concerned almost exclusively with affect as it is found in music—in its composition and performance. Starting with principles of Gestalt psychology, he has described in depth how expectations, their products and their violations, produce meaning and emotion in music.[31] Another eminent writer on the musical world, Charles Rosen, has also spoken about the role of discrepancies and discontinuities: "what is important is the periodic breaking of continuity. . . . The emotional force of the classical style is clearly bound up with [the] contrast between dramatic tension and stability. . . . I do not want to turn Haydn, Mozart, and Beethoven into Hegelians, but the simplest way to summarize classical form is as the symmetrical resolution of opposing forces."[32]

I cite these instances to draw attention to the fact that problems of value, in this case of aesthetic value, have not been unattended, and that when they are analyzed, phenomena of contradiction and discrepancy appear just as they do in the analysis of the social order.

The Social Context

Schemas that give rise to feelings of quiescence or familiarity on the one hand and contradictions on the other develop out of experience in the social context—they depend by definition on the experiences that people can have and have had in their social world. In addition to the schema-driven values that depend to a large extent on actual experiences in the world, social contexts frequently generate individual values independent of any experience with the valued object or event. A society and its members may reject as bad a variety of different people, objects, and events, even though the majority of its members have not had any contact with the offending object, and we also often accept as benign or good similarly inexperienced objects. This applies to social and racial prejudices as well as to the mundane world of food and clothing preferences. Americans "know" that communism, horses, and kilts are unacceptable to be espoused, eaten, or worn, just as members of other societies "know" that capitalism, pumpkin pie, and faded jeans are unacceptable. These values and beliefs are not necessarily directly "taught"; they are part and parcel of the society in which one grows up and lives. Nor are they learned by a neutral organism, but the very conditions of living at a particular time and place—in a specific historical and social milieu—determine what can and will be known and valued. These values are often adopted without any specific, concrete experiences with the liked or disliked event or object. However, they may be just as powerful in influencing the social experience of the individual—they represent the schemas that determine the cognitive structure of a particular society. How uniformly they are held depends to a large extent on the means by which they are promulgated. When the major means of communication and dissemination are primarily in the hands of a particular group or class or interest (as in the United States and much more so in totalitarian societies), they become unassailable parts of the society's structure. Cognitive contents not only are not context free, they reflect, incorporate, and constitute the society in which they are formed.[33]

Social (cultural) conditions also determine many value-laden actions and beliefs that are considered by some to be innately determined. I present in chapter

8 an analysis of aggression and hostility derived from a biologically determined condition—that is, reactions to thwarting—that makes its emergence as aggression dependent on social conditions. It is this kind of analysis of the proximal causes of complex values, such as hostility and violence, that I favor in assessing the interaction of biological factors (persistence, force, sympathetic nervous system reactions) and social ones (solutions to problems of thwarting).

Social values form their own schemas, and often the most stable ones. What social factors favor such stable values? We can assume that stable societies lead to stable systems of values. But such stable values may be also most vulnerable when they encounter serious discrepancies because the discrepancy has not been previously encountered—it is indeed novel and disruptive. There is extensive evidence in the psychological literature that experiential schemas that incorporate possible deviations and failures are most impervious to disruption.[34]

Serious challenges to value systems and ensuing difficulties in maintaining social stability occur when stable systems encounter inevitable contradictions. Social values develop from the interactions of people and groups of people. Such values may be competitive or cooperative: they may emphasize submission or dominance, depending on the social and historical matrix in which they develop. Only exceptional social (and personal) circumstances that generate unusual discrepancies and contradictions produce rapid changes in values and value systems. Such a gradualist view of values and changes in values is based on the assumption that the consciousness of values arises with and out of the structure of the society, which in turn depends on the slow change of groups of people acting in accord and forming their accepted value schemas.[35] Such change produces new values as well as their generational transmission. Only the most extreme utopians believe that values can be changed quickly by either instruction or revolution in the absence of the appropriate social climate.

Conclusion: Interdependence, Biology, and a Contradiction

A psychology of values does not and cannot generate a theory that provides a scientific account of what values should be. I have attempted to show that it can attempt to describe the conditions that generate values—without a commitment as to the content of those values. I do believe that an analysis of the regularities or consistencies and the inconsistencies or contradictions of human life and social institutions can inform our understanding of the creation and destruction of values.

My discussion of the various sources of values has been somewhat isolated from the interdependent context in which all these sources operate. Values are constrained by social and biological conditions, and in addition by the limitations that each of the sources of value places on the others. The sources of "external" constraint are biological, structural, and environmental. Biology constrains the kinds of actions we can undertake (because of our physique and biological constitution), and it also determines to some extent what will be considered pleasant or unpleasant. These biological constraints include the innate approach/avoidance tendencies and their derived actions that are perceived as positive or negative. The SNS reaction to discrepancy and the preference for the schematically acceptable event are also biologically given aspects of the human organism. One might conjecture that these factors determine the basic dimension of pleasantness/unpleasantness, the inherent conditions that make situations good or bad. Structural constraints are provided by the kinds of experiences that can or cannot co-occur, whether as a function of the limitations of the structural apparatus or as a function of the features and attributes available to the organism. Finally, the world we live in, with its ecology, meteorology, gravity, etc., determines the kind of schemas that can be developed and the kinds of structures that can occur. The biology of the organism constrains the kind of structures that can be formed and the kind of social organization that are possible, but the social organization also constrains how the innate reactions can be experienced or exercised.

None of the sources of value presumably ever produces an experienced or exercised value in its "pure" or unattenuated form. Innate sources inform structural and social formations, and those in turn modify the derived innate expressions. Values are mostly schema-dependent, and schemas are developed primarily in the social context.

I started this chapter as an attempt to discuss in a principled way the sources and conditions of values. I certainly do not claim completeness for this account, but I can hope that it will be the beginning of further studies of value. There is at least one good reason that the study of value has been avoided: It is difficult and harbors the seeds of its own doubts. How can one have a theory of values when that theory may be just another instance of contemporary and evanescent social values? The study of values harbors its own contradictions when it becomes a social version of Zeno's paradox: with every advance, we seem to be at best halfway toward our goal—never quite able to reach it. Societies always change, sometimes slowly, sometimes quite radically; the values that they generate are constantly in a state of flux. And these values, in turn, inform and constrain a psychology of values. Can a psychology of values arise out of a society that seems to inhibit the study of values; would such a psychology contradict the basis of its social existence? The social sciences are not, and probably cannot be, value free. In recognizing that constraint, we can try to face up to our particulate value systems and the values that our society and sciences have been bequeathed by social and historical conditions. In our understanding of our own value systems, we can strive for a better approach to a scientific understanding of values.[36] Once social scientists are beginning to understand the origins and uses of values, maybe they can strive to do what so horrified our positivistic and "pure science" ancestors—to search for and to justify the good and the ethical.

The Social Fabric

Aggression and Other
Human Characteristics

∞

Man's chief pleasure is society.
Sir John Davies

The Functions of Society

To what extent are we creatures of the society and the times in which we live—
the products of the fabric of society interacting with a biological matrix? The
notion that human beings and their thoughts are defined by their social contexts
has been argued from both the left and the right of the political spectrum. I present two examples.

Gottlieb argues from the left against two assumptions: that "[p]sychological
states are the key determinants of human behavior; [and that] individual psychology is best explained by reference to infantile experience."[1] He explains:

Psychology may appeal to the individual experiences of any given person to explain certain aspects of his/her adult behavior. But the task of social theory . . . is
the explanation of society as a totality. This task cannot be accomplished by simply adding up all the separate individual psychologies. Social theory seeks to com-
prehend how individuals are shaped by collective practices and institutions, as well
as how the psychological internalization of those institutions reproduces them. A
reliance on individual psychology as somehow more fundamental than collective
social relations will always be unable to explain historical change. For historical
change is a process that we enter into not as individuals, but as members of groups:
classes, genders, nations, races, ethnic groups, local communities. . . . The psycho-

logical individual is essentially a being without historical identity. As a complex of unconscious drives, needs and responses, we could exist anywhere and at any time.[2]

Compare this with the statement by a conservative German sociologist:

> The greatest error of individualistic psychology is the assumption that a *person* thinks. This leads to a continual search for the source of thought within the individual himself and for the reasons why he thinks in a particular way and not in any other. Theologians and philosophers contemplate this problem, even offer advice on how one ought to think. But this is a chain of errors. What actually thinks within a person is not the individual himself but his social community. The source of his thinking is not within himself but is to be found in his social environment and in the very social atmosphere he "breathes." His mind is structured, and necessarily so, under the influence of this ever-present social environment, and *he cannot think in any other way*.[3] (emphasis added)

Given certain characteristics of human beings, it is inevitable that given specific circumstances they will adopt one particular, and not another, mode of social structure. The adaptability of the species makes specialization likely, and at least in historical times specialization in turn suggests cooperative living with different specialists contributing to the welfare of the group. The relative vulnerability of human beings to predators (early and late in evolution), weather, and food shortages also contributes to choosing reciprocity and cooperation. These two general traits of reciprocity and cooperation transcend the limitations of the isolated human individual. They occur, in some form, in all societies. However, this does not mean that their universality needs be ascribed to some genetic factor. At present there is little available knowledge about such genetic factors, though I note below some social, peacemaking capacities of nonhuman primates. As far as social determinants of human cooperation are concerned, consider that similar ecological and survival problems have created similar morphologies with quite different genetic materials. In chapter 3 I noted how evolution generates similar solutions to similar problems. In the same way, similar social problems generate similar social solutions. An example can be found in the pyramids at Geza in Egypt and at Teotihuacan in Mexico, built by obviously independent cultures. However, if one wants to build monuments that reach close to a godlike sun, then pyramids are the obvious solutions in the absence of the technology to build skyscrapers or Eiffel towers. Similar environmental and social pressures may be responsible for human cooperation and reciprocity.

I shall illustrate the general nature of social determinants of some human traits using a detailed discussion of aggression as a springboard, primarily because it has probably been the most controversial aspect of "human nature."

Aggression

Probably the single most debated social characteristic of humans our tendency to perform acts of aggression, ascribed by some to our biology and by others to our society. We start with the difficulty of defining aggression in any way that is satisfactory to all concerned, including those holding opposing views on the question of "innateness." There is a degree of agreement in the use of the term in the natural language which provides some consensus. Excluding the problems of aggression toward nonhumans, and I surely do not wish to enter into questions of animal rights, there is some sense in which we are all talking about any action by human beings that results in physical or psychic injury to another human being, or which intends such injury.

In order to contrast personal and group aggression, consider the case of intergroup aggression—wars as they are called by "advanced" societies—where aggression against others need not imply any intention to injure. The best outcome following an ultimatum or a declaration of war would be, in the interests of the aggressor, the immediate capitulation of the "enemy" without a shot being fired or a hair being bent. And that may well be the intention of the aggressor. However, the willingness to consider injurious action constitutes the aggressive action in this case. Up to this point, similar situations exist in intragroup situations where people are bullied or threatened to capitulate. But in the intergroup situation, it has generally been agreed by most writers on the topic that the individual participants need not intend to injure nor even be willing to inflict injury. The soldiers of aggressive armies are frequently not individually willing to injure, nor do they often intend to inflict injury. Even the directors of those armies—the political initiators—frequently give little thought to the problem of injury. Their concern is with conquest and victory, not individual aggression. On the other hand, I think it can be agreed that any attack on another group, or any willingness to injure another group, constitutes aggressive actions, regardless of how the participants "feel" or "intend." Thus actions that are aggressive do not permit us to conclude that intention or hostility was present in the actors.

Conversely, however, I would argue that the intention to injure or the expression of hostility, at the individual level, does constitute aggression in a generally accepted sense. That is, at the intragroup level, at the point of individual aggression toward other members of a group (however small or large), we need to consider the origins and conditions of intentions to injure as well as the actions. Understand, that I am not arguing that only intentional aggression is of interest at the individual level. Too often, individuals engage in aggressive and injurious actions without experiencing hostility or the intention to injure. What I am saying is that we must understand both such "unconscious" aggression and conscious feelings of hostility.

In the context of the discussion of emotions in chapter 6, what constitutes the "emotional" experience of hostility? Apart from the arousal, the values or evaluations that constitute such feelings are associated with injury to another, destruction of another's ability to act in some way, or a sense of revenge—that is, the need to punish. Most of these are associated with the conditions that give rise to arousal in the case of aggressive or hostile feelings. The arousal is usually associated with the perceived or actual case of another person blocking some need, desire, want, or relationship. In Western society when another takes your job, deprives you of your loved one (lover, child, or friend), or even your opportunity to get onto a bus or to get to a traffic light before they do, these interruptions are occasions for evaluations that have at their core some "injury" to the other person.

Types of Aggression

Before discussing problems of aggression and hostility in detail, we need to take account of categories of aggression—hurtful actions against others, which seem to be different from the paradigm cases, in which there seems to be some degree of justification. Primary among these are defensive and parental aggression.[4]

DEFENSIVE AGGRESSION.   The most frequent defense against accusations of aggressive acts is the argument that the acts were necessary in order to protect oneself, in order to avoid injury or even death. It makes biological sense that the organism will act to avoid its own demise—the selfish gene argument comes to the fore here. People do and should protect themselves to avoid being killed. Some pacifists have, however, taken even this position to be inhuman and indefensible—any act that hurts another human being is considered immoral. Rare as such attitudes are, they do illustrate how a cognitively complex organism like a human can overcome what seem to be fairly fixed biological traits.

Yet one must beware of the equally cognitively complex argument that uses the tendency to protect oneself in order to justify aggression against others. How many wars have been fought under the banner of defense against an aggressor— a justification frequently used by both sides? The understandable need to aggress against the person who is, here and now, threatening to injure me cannot be used as a basis for trying to inflict injury on other individuals just because they are members of the enemy's army, society, or nation. The people in "enemy" cities, the soldiers in the "enemy" army are not proper objects for defensive aggression. When defensive justifications of aggressive acts are used to justify aggression against the weak and unprotected, then we are no longer dealing with defensive aggression. If wars are to be justified, they must be justified on grounds other than the need for defensive aggression.

PARENTAL AGGRESSION.   Another class of aggressive acts is motivated by a desire to protect one's child. These acts are usually carried out by parents or caregivers. Such aggression is designed only to keep one's child from harm. The category can be seen to include two kinds of aggressive acts: toward the child and toward an outsider. The former is designed "for the good of the child." A mother may slap a child who is dangerously close to a pot of boiling water. Except for cases of child abuse, where the aggressive act expresses some need of the parent that is unrelated to the need of the child, these acts have no aggressive intent in the sense of wanting to injure or debilitate the child. There is, of course, a fine line between protective "punishment" of the child and expressions of hostility (and displaced anger) by the parent.[5] But a case can be made about the essentially benign nature of some cases of parental aggression, though they do interact with complex questions of the "rights" of parents and children.

The other kind of parental aggression is directed against an outsider who is seen to harm, or potentially harm, the child. This is really another case of defensive aggression. If we consider the child as an extension of the parents' ego—that is, the parents consider the children as part of themselves, then the parent is really defending him or herself against an aggressor who is potentially likely to harm the self. Here again, one needs to be careful about cases where parental aggression is used as an excuse for a blatantly aggressive act. The argument that one is participating in a war in order to defend the child against the horrible "hun" or "red" or "Jap" hides the justification for warfare that has little to do with parental defense—in very few cases (the Holocaust being an exemplary exception) does the "enemy" have any designs on the children of the opponent. And note that the argument has often been used symmetrically in the past—both sides are seen as threatening children.

The Innateness of Aggression

I want to discuss three positions that have dominated the argument for the innate nature of aggression. The *drive argument* is exemplified by the work of Lorenz. The argument postulates, primarily for mammals, innate aggressive drives that are usually released in interspecies competition. With the dominance of the human species, the opportunity for such aggression has significantly decreased, though it is still found in hunting sports and similar activities. Instead of interspecies aggression, humans are said to have turned to intraspecies aggression, exemplified in wars and group conflicts in general. Implied in the argument is that there exist innate releasing-conditions that lead to the activation of the aggressive drive.[6] It is generally difficult to define what these conditions might be, though it is not difficult to define the social conditions that often lead to intraspecies aggression between subgroups of humans. It has been argued that aversive conditions are the major releasers for aggression. But humans are cognitively complex enough to have developed a large variety of ways of dealing with aversive conditions and situations. As an argument for explaining the occurrence of wars, the drive argument is rather weak. Certainly at the phenomenal level, most soldiers involved in a modern war do not experience hostility or aggression against the enemy—they would mostly rather go home. Aggressive drives do not describe the reasons soldiers go to war.[7]

The *adaptiveness of aggression* is another argument that has been advanced for the innateness of aggression. The occurrence of territoriality is produced as one of the major reasons for aggressions—it is an attempt to defend the territory, the ecological niche, of the organism, which is necessary for its survival. But territoriality in animals occurs for a variety of different reasons and under a variety of different conditions. Territoriality may protect food supply, family, offspring, etc.—all of which lead to reproductive success. However, there is evidence that much of human intergroup aggression has little to do with territoriality as it is found in other animals. It is often related to expansionist, power-related motives that seem to have little or no connection with territoriality, except for symbolic invocations of territoriality that serve power needs.

The *evolution of aggression* invokes common primate *evolutionary bases of interpersonal aggression* and has been advanced by some biological anthropologists. The argument is that some behavior patterns are shared by all primates and that knowledge about primate behavior can be useful for an understanding of human behavior. Even though our common ancestor with our closest primate relative, the chimpanzee, dates back some 10 to 15 million years, there is some utility to such an approach. If one wishes to use primate behavior as an initial model for human traits, the question arises: Which primates and which behavior?

Aggression and Our Primate Cousins

Baboon societies have been a rich source of cross-species insights and have been used in particular to paint a picture of a society dominated by big and powerful males who fight each other for limited resources of food and mates. The successful aggressive baboon then gains the appropriate reproductive advantages. The main gain of a competitive aggressive strategy is the resultant dominance hierarchy among the males (including dominance by any male over all females) and the relative advantages in access to food and mates. In addition, the geographical structure of the baboon troops also suggests the use of the hierarchically arranged males in fighting and defending against predators.

More recent studies of Savannah baboons by Shirley Strum have undermined much of this argument. Savannah baboons are one of a small set of primates (humans being another one) who have moved from a forest environment into open areas. Thus, the new environment may bring with it new options for the management of social arrangements, just as the development of language equipped complexly cognitive humans with alternatives to dominance and aggression.[8] Strum found a baboon society in which there was much less aggression than one would expect on the basis of other reports of baboons and primates in general. Social relations showed a relative absence of male dominance hierarchies and a prevalence of "friendships," particularly between male and female animals. In a situation where aggression may turn out to be costly, intelligence and social skill come into force. The family structure of these animals is quasi-matriarchal, with pronounced female hierarchies both within a family and within the troop. Social play by the young begins very early, as do friendships of youngsters who play together. Some mothers form friendships, but most interesting are the male–female friendships. The relationships are reciprocal and are based on trust, and they are more often between biologically unrelated animals. One result of this structure is that there is no obvious male dominance hierarchy, and aggression, when displayed, more often fails to get the prize of a mate or food than being successful. Males often use agonistic buffering (interposing a baby or female) to discourage potential aggressors. Aggression takes place when newcomers join a group, since males usually leave their troop at adolescence. The newcomer will frequently try aggressive intrusions, but this usually fails, and it is only when he has established stable friendships that he slowly becomes accepted and sexually and reproductively successful. In the end, we are presented with a new model for male success: finesse and social strategies replace aggression and competition as successful adaptive strategies.

Another example that differs from the popular notion of naked aggression among animals other than humans comes from the observations of Frans de Waal (1989). His observations demonstrated the existence and practical inevitability of reconciliation or peacemaking, both after aggression has occurred or instead of overt aggressive action. The intricacies of these strategies have been documented with the behavior of several different primate species (including humans). For example, among the bonobos (also known as pigmy chimpanzees and closest to humans genetically and apparently also intellectually) sexual play is used as much for the relief of tension and reconciliation as it is for reproductive purposes. Sexual and other physical contact tends to increase after aggressive or other tense interactions. Thus, one bonobo, after biting the toe of another bonobo who had temporarily carried off her child, returned shortly afterward to lick the blood off his injured toe. De Waal's conclusions generalize the social strategies observed by Strum in baboon society. He concludes that the emphasis on competition among animals (including humans) is a "dreadful simplification. Antagonists do more than estimate their chances of winning before they engage in a fight; they also take into account how much they need their opponent. . . . [I]f aggression does occur, both parties may hurry to repair the damage. . . . [and] appropriate countermeasures evolved along with aggressive behavior, and . . . both humans and other primates apply these measures with great skill."[9]

If aggression is not inevitable for baboons or chimpanzees, it cannot be an inevitable consequence of our primate inheritance as humans. Presumably, modern as well as early humans were not less socially intelligent than our primate cousins. Social negotiations, reconciliation, and peacemaking, together with resources such as language and tools, are an alternative to an ideology of aggression and competition. With this new evidence, aggression cannot anymore be seen as inevitable (nor as effective) as it has been previously believed to be. Aggression is obviously a remaining option, but now it is only an option.

Finally, the notion that aggressive behavior is adaptive in the cause of protecting oneself and one's goods against predation of whatever kind is somewhat circular. Obviously, aggressive behavior would be neither necessary nor necessarily adaptive in a niche in which no aggression occurs. If nobody threatens one's goods, then there is no need to aggress against "threatening" conspecifics. Thus, the argument for the adaptive nature of aggression assumes, in part, the presence of aggression in the first place.

Before advancing an alternative argument for the biological origin of aggression when it does occur, I need to note that aggression is not the only and sometimes not the best, nor the preferred, way of maintaining reproductive success. The alternative lies in flight, in the avoidance of conflict and threat to one's existence, family, or offspring. Running away is a widely used method of avoiding conflict and injury. The avoidance of conflict is a characteristic of some human societies, though it has been a socially devalued aspect in Western society.

A Basis of Human Aggression

In the case of aggression, I wish to make a specific point about possib
tionary mechanisms. The ease with which aggression seems to appear
ety of different societies and human conditions makes it plausible that some innate disposition makes the acquisition of aggression relatively easy or likely. On the other hand, both inter- and intragroup aggression are not a necessary outcome of human social interactions and human societies. Thus, I do not adopt a completely environmentalist point of view toward the phenomena of aggression, nor do I claim that there are not other ways in which aggressive behavior arises and is mediated. Instead, I ask what are the conditions that seem to play into our innate dispositions that facilitate aggression (and competitiveness, hostility, and other accompanying symptoms)? And one can then ask the equally important questions about the conditions that generate nonviolent, cooperative, sharing, and loving human societies.

I base my argument on three general dispositions: (1) arousal, the occurrence of sympathetic nervous system (SNS) arousal when an action is blocked or interrupted; (2) persistence, the tendency to repeat unsuccessful actions; and (3) intensification, the tendency to intensify unsuccessful actions—that is, to repeat them with greater strength or force. Of these, arousal is surely an innate characteristic, and the other two may well be or are based on relatively simple innate mechanisms. The occurrence of aggression and the contemporaneous or subsequent development of hostile, aggressive emotions, depends on occasions when actions are thwarted, blocked, or stopped, followed by arousal and by persistence and intensification of actions. I argue that the original development of aggressive tendencies, based on these three factors, takes place in very early childhood.

Consider the case of aggression, hostility, and anger and their possible development de novo in young humans in competitive, aggressive societies. Given an infant who is thwarted from reaching some goal or object, we would expect that this blocking of a goal would generate SNS activity. In other words, the interruption or difference detector would be activated in this, as in any case of frustration or blocking. The child may then continue to try to reach the object and may, under most circumstances, persist in its efforts and also intensify its actions toward that object. The result of such repetition and intensification is that the blocking agent—the object (frequently human) that stands in the child's way—is likely to be struck or hit as the infant moves toward the desired object. If such action is successful, it will be repeated for that reason, in the presence of SNS arousal. In order to have such actions against blocking agents to become part of the child's repertory, it is necessary to have social approval for such actions. Thus, if the action against the (human) blocking agent is valued positively by the social environment (or not valued negatively), the generalized "aggressive" behavior will be repeated on other occasions. In other words, the social context (in this case the caretakers) determines whether these actions will be reinforced. Eventually, the aggressive action becomes mentally represented with a particular value and can emerge as a relevant image at the occasion of thwarting. That concatenation of image, thwarting, arousal, and reaction then represents one outcome of the child's emotional development and generates hostile, thoughts when thwarted and the hostile, competitive reaction to thwarting.[10]

There is some cross-cultural evidence that the aggressive response to thwarting is not "built in." Different kinds of childrearing can produce children and adults who learn different behaviors when thwarted, such as acceptance or co-operation, and who do not become angry when frustrated. They in turn constitute a cultural environment that does not encourage the early intensification of reactions—the aggressive or hostile response to being thwarted. On the other hand, if predispositions toward violence exist due to early childhood experiences such as rearing techniques, then the question arises under what conditions violent aggression is more or less likely to happen. It seems clear that conditions of social stress, frustration, discrimination, economic deprivation, and other social conditions will lead to increases in violence.[11]

The discrepancy/difference detection system is one avenue of producing aggressive behavior, but that does not mean that this system is the only evolutionary road to aggression. Some of the types of aggression (e.g., parental) discussed earlier may require a quite different selection history. Nor should it be assumed that the discrepancy/difference mechanism is the only source of arousal.[12]

I now turn to some illustrative examples of social organizations that are built without extensive recourse to aggressive or competitive action.

A Look at Alternative Social Organizations

The Semai

The Semai of Malaysia have constructed a nonviolent society that approaches aggression not by forbidding it, but by denying its existence.[13] Their attitude to anger is not that it is bad, but rather that they just do not get angry. They believe humans to be basically nonviolent and nonaggressive, and they employ very few direct expressions of aggression. Quarrels occur, but they make both the participants and observers uneasy. Interestingly, when intoxicated the Semai become talkative and noisy, rather than aggressive.[14]

Semai children are raised with the understanding that children learn by themselves. The crying child will be typically left to cry—to cope on its own. Physical punishment, when used, is superficial, consisting of pinching children's cheeks or patting their hands. Children are very early introduced to the concept of *bood,* the notion that instead of rebelling one can simply *bood,*—that is, not feel like doing something or other. This is an acceptable response to a request or order and is not punished. If violence occurs in child's play, it is not reacted to with violence. Aggression is not punished, but countered with vague threats of spirits that would punish them. Essentially, discipline is enforced by scaring children, usually with threats that are rarely, if ever, carried out. Aggression by children is typically countered with laughter or threats. It is important to note that the would-be aggressive child has no, or very few, role models. In addition, and in keeping with the general trend of childrearing, there are few competitive games among children.

The Tasaday

Extensive material is available on the life and culture of the Tasaday of Mindanao in the Philippines. Unfortunately, there also exists an acrimonious debate on whether the Tasaday are a genuine isolated tribe, emerging from a stone age society, or whether they are a tremendous hoax engineered by former President Marcos' underlings. The answer is probably somewhere in the middle, in that the Tasaday were neither isolated for hundreds of years nor a hoax, but a relatively circumscribed small band that developed much of its own culture, but still part of the surrounding cultural environment, and that they were subsequently exploited by the Marcos regime. I present whatever evidence there is with the proviso that the reader might want to use a reasonable amount of caution in accepting it.[15]

The Tasaday are a small group of about 25 men, women, and children, who live in the rain forest in the mountains of Mindanao in the Philippines. They are cave dwellers, with a stone tool technology and a gathering economy. It is believed that the Tasaday separated from their parent group over 500 years ago, and there exists some substantiating archaeological as well as linguistic evidence.[16] The first regular contact with the dominant surrounding culture was made in the 1960s. I concentrate on those accounts that describe their life in aspects uncontaminated by their recent contact.[17]

The Tasaday were nomadic gatherers and foragers, who engaged in some hunting, which was limited by the absence of weapons. Apart from the recent contacts, their major interaction with other groups is claimed to have been with two other, similar groups in the rain forest. However, their contacts with those groups play an important role, because wives were apparently "obtained" from those groups, and young Tasaday women may join those tribes. As far as is known, these contacts were always friendly; wives were not "stolen" but may have been "bought" for food and implements.

The single most striking aspect of the Tasaday was their noncompetitive mode of relating to one another. They did not seem to live *strongly individualistic* lives. After the initial contacts with journalists, officials, and anthropologists who visited them, they quickly formed enduring affectionate ties with many of these visitors. Observers have been unable to note any quarrels; fighting seems unknown among them. Hostile, aggressive behavior among adults was unknown, but joking and teasing took place frequently. There are no weapons, either for interpersonal use, or even for hunting larger animals. They seem to have few fears, the exceptions being thunder and snakes. The Tasaday shy away from people who are loud, and they have "less love" for people who have "sharp looks and loud voices."

In a society that seemed to be faced with two major problems—adequate food supplies and the availability of wives—the solutions were represented in a continuing commitment to food sharing on the one hand and strict monogamy and sexual noncompetitiveness on the other.

The reconstruction of the Tasaday's level and mode of subsistence shows a people who have their wants "easily satisfied by desiring relatively little, by being satisfied with a lower-than-normal dietary level with its attendant lesser energy expenditure (or greater leisure) and largely unelaborated material culture."[18]

Food sharing, sometimes described as "meticulous," has been noted by all observers. The products of foraging are shared among all members of the community, and portions are left for those who might be absent at the time the food is consumed.

Children receive love and affection from all members of the group—they are made to feel important. Young children are always the first to be fed. Children react to thwarting and frustration. They cry when hungry; they show stranger anxiety. However, the parent's reaction is one of acceptance, not of punishment. The children do show competitiveness and aggression; they will struggle over food and toys and compete for attention. However, by the time they enter adolescence, these traits—in terms of interpersonal relationships—have dissipated. Little is known about the Tasadays' specific means of handling the children's reactions to frustration, except that when these responses are evident, the child will typically be distracted, given attention, and firmly but gently disciplined. Children are never slapped or struck in the process.

Social Alternatives: Aggression and Cooperation

In order to evaluate the generality of human aggressiveness, we need careful cross-cultural analyses which are not based on supposition or ad hoc arguments. Citing the Tasaday children's rivalries, the ethologist Iraeneus Eibl-Eibesfeldt saw evidence of aggression and proof that it is a basic human characteristic.[19] But ethnocentric interpretations of isolated behaviors cannot be substituted for careful analyses.

Among the more sophisticated mythmakers about necessary human aggression, C. J. Lumsden and E. O. Wilson have arrived at definitive conclusions about preliterate human groups. For example, they note that "[m]ost present-day groups of primitive men [sic] engage in some form of organized aggression" and conclude from that "that ritualized fighting and war were also widely practiced by our ancestors."[20] "Most" is an inadequate quantifier for making evolutionary arguments, and at the very least one would have to exclude the Semai and the Tasaday from "most." But Lumsden and Wilson are more specific. They refer to the "evidence of violent aggression among recent bands of hunter-gatherers, whose social organization most closely resembles that of primitive man."[21] And they adopt a hypothesis of Darwin's that war strengthens the intellect,[22] by asserting that a band (of hunter-gatherers) would consider how to deal with adjacent social groups in an "intelligent" fashion: "A band might then dispose of a neighboring band, appropriate its territory, and increase its own genetic representation," and so on.[23] They conclude that "[a]ggressiveness may well be the dark underside of the human intellect."[24] At least Lumsden and Wilson agree that aggressiveness cannot be identified as the prime mover toward human culture. But their tacit agreement with nineteenth-century myths about the evolution of human aggression clearly does not apply to all societies. Maybe these societies are not very "intelligent." But these "primitive" people surely require us to think again about primitive reconstructions of imagined behavior of our ancestors.

The "primitive" society favored by the advocates of an aggressive humanity is that of the Yanomamo who live in the rainforest of Venezuela and Brazil. Until comparatively recently, their primary economic activity had been hunting and gathering. They live in small groups of 35 to 200 persons, at a distance of about a day's travel from the nearest Yanomamo group. These groups engage in repeated warring raids on one another, and the result is injury and death. Over half of the male deaths are due to these "wars." In contrast to other groups, who are also "short of women," the major cause of the fighting is the acquisition of women. The similarity in "primitiveness" of nonaggressive groups and the Yanomamo, both in history and in their economic activity, contrasts with the extremes in intragroup (and extragroup) aggressiveness that these two societies represent. The dissimilarity also extends to their mate selection patterns. Napoleon Chagnon has shown that the cross-cousin marriages favored by the Yanomamo provide an excellent illustration of kin selection theory with particular reference to inclusive fitness.[25] In contrast, some societies seem to avoid within-group marriages, though cross-cousin selection would be relatively easy, and instead tend to find their mates from neighboring similar tribes with whom they generally have minimal contact. Thus, kin selection theory can be supported with evidence from one group but fails with another.[26]

The contrast between peaceful and aggressive societies should not suggest that they are extremes. We do not have too much evidence about the cooperative/competitive dimension in early hunter-gatherer societies, which represent most of the history of *Homo sapiens*. What evidence there is for pristine hunter-gatherers is that they are primarily cooperative and egalitarian—and that the Yanomamo are the exception. A seventeenth-century report on the Montagnais of the Labrador Peninsula supports this argument.[27] The talent of the individual was recognized as benefiting the group as a whole. Paternity of children was considered relatively unimportant—the children were "of the tribe." Necessities and staples were shared, and both women and men participated in their production. Similarly, among some American Indian groups, cooperative work in the fields was the rule.[28]

Even a cursory study of human societies convinces us of the varied and different ways in which human beings can organize their need for social interaction and interdependence. Having seen some examples of such diversity, I now turn to a different kind of diversity—the differences among individuals.

Differences Among People

Intelligence and Gender

∞

We live in deeds, not years; in thoughts, not breaths,
In feelings, not in figures on a dial.
                                        P. J. Bailey

The Problem of Intelligence

Probably the most frequent answer to the question of what distinguishes humans from other animals is "We are smarter, more intelligent." But what is implied by that answer? In part, it is a simple statement of the fact that humans have dominion over the animal world. If the other animals were brighter, they would not let us do what we do to them. But what makes for this obvious difference?

The answer is clearly multidimensional—in the first place we have language and a transmitted culture. Humans accumulate and transmit knowledge over time. And even though written language is probably no more than 6,000 years old, the oral traditions enable the adequate transmission of cultural knowledge. Second, humans are exquisite tool users—an ability that has engendered our modern technology. Tool use depends to some extent on the evolution of the upright primate—freeing the hands for tool work and for further evolution in dexterity.

Finally, the evolution of the human brain made possible the conceptual apparatus that our knowledge-handling brains have developed. Logical thought, generalization, and categorization, the proliferation of a vast network of parallel systems funneled through consciousness, all make for a problem-solving animal of great skill, although our abilities are probably in their early stages, given the short time since the human brain reached its present size and complexity. We might take note, though, that other animals—for example, cockroaches and whales—are also very intelligent in the sense that they are, except for human encroachments, exquisitely adapted and attuned to their habitats, their ecological niches. Not as much can be said of the habitats some humans have created—slums, poisoned waters, and deforested areas, among others.

So much for the difference between humans and other animals. But that is not enough for the products of Western societies. We apparently need to know how bright each one of us is and how much brighter some of us are than others. The competitive spirit has dimmed (or illuminated) the entire question of intelligence. The past century in particular has witnessed an increasing and overwhelming concern with measuring the "intelligence" of people and differences in "intelligence" among groups of people. It is in the latter context that arguments have arisen about evolutionary differences between groups and genetic differences among individuals. The primary weapon in this assault has been the development of the intelligence quotient, the IQ. Given that the IQ was developed initially by Alfred Binet in order to assist the French school system in identifying children who might need special attention to bring them up to the level of their peers, its subsequent misuse is even more astounding. Binet wanted to know whether children performed, in school-related tasks, at, below, or above the performance of their age peers. And voilà—the IQ, the ratio between mental and chronological age was born.

The sophisticated intelligence tests of today still are targeted at education-related skills. IQ correlates consistently with years of education and, more important, with educationally and intellectually sophisticated home and school environments. The common wisdom is that you are born with a particular dollop of intelligence—forever damned or elevated by the size of the dollop. This point of view has been fostered by elites—those people who are likely to test high in intelligence tests—and by the people who generate and use intelligence tests, the testing industry, and some educational establishments. The single-score IQ is still dominant, despite patent indications that individuals vary widely in specific and often unique skills and knowledges, and despite valiant efforts to demonstrate and use multidimensional tests of intelligence.[1] The IQ score is, for all intents and purposes, the symbol of intelligence writ large.[2]

The most noticeable social effect of this view has been the assertion of genetic differences among human groups. This position speaks mostly to presumed white and black differences in the United States, but it has also in the past been applied to immigrant versus native groups in the United States, different ethnic groups in India and Japan, upper and lower classes in nineteenth-century Europe, and many more societies.

It is a fact that tests of intelligence show a more or less consistent difference of about 15 IQ points between the American black and white populations. I do not want to spend much time on the statistical arguments that this is genetic. Others have critically addressed the data that purported to show such genetic factors.[3] I am more concerned with the data that suggest a different answer to the puzzle of black and white differences, and which I discuss later in this chapter.

The pursuit of racial differences has also given new impetus to the desire to find the genetic basis of intelligence in general. Whereas work on the relation between genetics and intelligence is usually carried out under the banner of disinterested "pure" science (and I assume that some of the workers in the field are sincerely committed to such an attitude), much of this work feeds back into questions of race. Thus, if intelligence differences are mostly genetic, and if there are consistent differences between minority or indigenous and white populations, then it seems to many quite reasonable to assign inferior genetic endowment to the minority populations, whether they are African American, New Zealand Maori, Australian Aboriginal, or Indian lower castes. This despite the fact that both anthropologists and population biologists have repeatedly noted that the concept of "race" is a historical, constructed category, rather than a genetic, scientific one.

One of the most puzzling aspects of this kind of research is that it is usually concerned with "single-number" intelligence—with assigning genetic causation to the single summary measure of intelligence obtained by the usual tests. The illogic of such a pursuit is demonstrated by the fact that people differ in their "intelligence" as demonstrated by different skills, tasks, and situations—some are good at reading, others at toolmaking, others at arithmetic, others at interpersonal sensitivity, others at ballet, others at baseball, others at painting, and so forth. The more sophisticated tests do, in fact, provide subscores in two or more areas, which demonstrate directly the measurable differences in skills between and within individual performances. How such a varied and typically human picture of "intelligence" can be reduced to a single number or how that number could then be directly related to genetic factors is puzzling indeed.[4]

Since it has been argued that the best test of genetic influences on intelligence is the investigation of the scores of identical twins, separated at birth and reared in different environments, I need to extend further the discussion of twin research presented in chapter 3. To summarize, one of the claims of this work has been that it is possible to separate the relative contributions of nature and nurture by holding genetics constant (with identical twins sharing identical genetic backgrounds) while varying the environment (when identical twins are reared separately). As I noted in chapter 3, the only proper test of that relationship would require (1) the assignment of the twins to adoptive environments chosen randomly from the full range of possible environments, (2) evidence that these adopted children are treated, educated, and schooled no different than are other children, (3) measurements of environmental variables that are as valid and reliable as the measurement of genetic and test variables, and ideally (4) an (impossible) random assignment of genetic sampling of the twin population used to assure the generality of one's findings.

To illustrate some of these points, one of the most extensive twin studies uses adoptive father's and mother's education and the father's socioeconomic status (SES) as well as physical facilities in the homes as indices of environmental similarities.[5] These are at best rather superficial measures, particularly if one is interested in the degree of stimulation, interaction, encouragement, etc., to which the children are exposed. Nor do we have any normative, taxonomic information about these variables; we do not know how they are distributed within and between cultures. However, more important might be the fact that even some of these superficial measures are significantly correlated between the two adoptive homes—that is, for mother's education, father's SES, material possessions, and "mechanical" facilities.[6] In other words, the two environments are similar, they are correlated, as is the genetic endowment of the twins. Given those data, little can be said about genetic and environmental distinctions. One would like to see a more sensitive measure of the sociocultural environment, including some composite measure of the available social information as long as single measures are used for "intelligence." Until such time when the adoption problem has been solved, when assessment of children's current and past environments can be accurately and precisely measured, it is best to suspend any judgments about the genetic basis of intelligence.

The Social Basis of Intelligence

My main purpose here is to demonstrate the importance of the social milieu, the influence of the immediate social setting on intelligence scores. The purpose is not to prolong further the use or defense of such scores, but rather to show that tests that address the ability of individuals to cope with schools are directly related to attitudes, expectations, and skills acquired in and appropriate to the individual's social setting. In particular, I examine some of the evidence that differences and changes in such social factors have direct and measurable effects on intelligence test scores.

The most obvious changes in social conditions occur over time. Societies change rapidly, particularly in recent times, and so do the social factors that determine attitudes toward intellectual activities, define the skills required and the technologies that support such skills, and generate expectations toward schools and intellectual activities. Flynn has shown that intelligence scores have gone up some 10 to 20 points in a 30-year period from the 1950s to the 1980s in most Western countries, such as France, Holland, Norway, and the United States. If IQ is truly genetic, then there would have had to be a fundamental genetic change in these countries, which, of course, there was not and could not have been.[7] Another way of looking at this change tells us that contemporary populations score significantly above the average if they take IQ tests from some 30 to 50 years ago, scored against the norms of those days. Thus, this increment suggests that some unspecified social difference can make important differences in measured intelligence. If the changes in these societies can, in 30 years, produce a difference in IQ that is about the same as the difference between black and white populations in the United States, then it seems reasonable to assume that similar social changes for the black population would produce IQ scores similar to those of the white population. In addition, the data that Flynn summarizes show that there was a shift of about 15 IQ points in both black and white Americans between 1930 and 1980. This finding indicates that the white U.S. population in 1930 had the same IQ on the average as the black population in 1980—but where is the difference in genetic basis for these two white populations? Put in another way, scored against current norms, the white population of the United States had a mean IQ of 85, but they are the very population that developed our modern industrial civilization.[8] Some aspects of the changing environment have had substantial effects on IQ scores. Blacks show the same effect of a changing environment, and one would expect that their average IQ would be much higher if the environment of educationally disadvantaged children were to change.

The basic problem is one of comparing "actual" intelligence and IQ, since it is unlikely that, for example, the estimated IQ of 45 of the British population of 1892 or of an IQ of 79 for the U.S. population in 1918 represented the practical and theoretical intelligence of those populations in tackling the problems of their world.[9] Up to now I have just globally assigned these effects to the surrounding social context. Specific theories have been proposed to account for these gains. One of them is the likelihood that people spend more time in school and more people are exposed to educational opportunities. However, data support this hypothesis only in some, but not all, cases. In the United States, for example, increases in IQ occurred at a time when there were no or little academic achievement gains. In addition, changes in the content or the complexity of the social environment as well as changes in nutrition (and resultant brain size) have been advanced as explanations of these effects. Changes in environment have been invoked that are supposed to affect the IQ of all members of a society equally.[10] But these beg the question about the innate nature of intelligence. If IQ is taken as a measure of intelligence (as is customary), and if it is as sensitive to environmental change and variation as these findings and explanations indicate, then IQ is a poor measure of underlying (native) ability and much better suited as a measure of environmental variations and differences. As Flynn has noted, we may someday have adequate data that tell us why different tests seem to behave differently for different populations and for different times, and the result will be a better theory of intelligence.[11]

"Race," Caste, and Intelligence

What are the social causes of the sharp differences in IQ and in school performance between minority and majority populations, not just in America but also in other cultures?[12] Without doubt, black populations in the United States are exposed to social differences in terms of family environments, information and media facilties, formal schooling, and family attitudes toward education. John Ogbu of the University of California at Berkeley has pinpointed castelike attitudes toward education and schooling which have "turned off" the interest in education and the ability to break out of the minority environment.[13] The inadequate school performance (and by implication at least, the lower IQ scores) of black and other castelike minority populations is a symptom of the way these minorities adapt to their occupational and social status in the life of the majority community. The barriers to occupation and status that their castelike status erects affect attitudes toward schools and influence the way education and academic achievement is treated by and within the caste population. The skills and competence required within a caste demand different cognitive and linguistic skills from the unattainable and barred technical and social achievements that are open to the majority class. Ogbu's central hypothesis is that "black school performance is an adaptation."[14] As far as the American environment is concerned, he concludes that home and school are not the ultimate source of deteriorating school performance; instead, they are the "media through which the influences of the caste system are transmitted to the black child."[15]

Ogbu surveys other castelike groups as well. None of these minority groups (whether in the United States or Japan or New Zealand) is economically in the mainstream of society, and their job ceilings and opportunities are low. These factors engender an attitude of defeatism toward any possibility of going beyond the achievements of the caste as it is currently constituted. Schooling is not seen as a realistic avenue to a better social or economic status, and the castelike status of the minority groups perpetuates these attitudes as the normal ways of looking at possibilities and potentials. In addition, teachers expect members of these "castes" to perform more poorly than members of the majority and thus tend to treat them in ways that make these expectations come true. In fact, there are indications that children from minority groups who do try to emulate the educational goals and attitudes of the majority population are often seen as "selling out."

Among the castelike populations that Ogbu surveys, in addition to African Americans, are West Indian immigrants in Britain, native Maoris in New Zealand, scheduled castes in India such as the "untouchables," the Buraku of Japan (racially no different from the dominant population, but considered as inferior and relegated to menial occupations), and oriental Jews in Israel (immigrants from the former Ottoman empire). When available, the IQ or comparable scores of these minority groups are lower than those of the majority class, and interestingly the difference is about the same as that for blacks and whites in the United States—for example, it is approximately 13 IQ points for the Maoris, 15 for the Burakumin, and 13 for Oriental Jewish children. The characteristics of these people are described in terms similar to the derogatory epithets aimed at the African-American population. The Burakumin are "mentally inferior, incapable of high moral behavior, aggressive, impulsive, and lacking in any notion of sanitation or manners."[16] The Maori are "lazy, happy-go-lucky, undependable, and capable of only rough manual labor."[17] All of these populations have in common lack of access to the occupations and academic facilities of the majority population, and all have developed a caste culture that adapts to these barriers and restrictions.

It is of some interest for those concerned with the genetics of intelligence that the difference in performance and IQ tend to disappear when members of these minorities break out of the caste environments. In particular, it should be noted that within the caste environment, minorities of the same genetic endowment as the majority class perform at the low level, but once outside this environment seem no different from their majority brothers and sisters. Thus, Oriental Jewish children, transplanted to Israel, whose fathers attain higher education are essentially indistinguishable in achievement scores from the average of all children with fathers of the same educational status. Burakumin children whose families have emigrated to the United States show no difference in achievement from non-outcaste Japanese children in American schools.[18]

This kind of approach produces a different answer from the usual statistical studies of intelligence test scores, twins, and racial groups. By presumably holding genetic endowment constant and varying the social environment, we can find large differences in achievement or intelligence. But in general, it seems premature to talk about "intelligence"—derived typically from complex intelligence and achievement tests. Once again, we are faced with complex, highly intricate human behavior and the attempt to find simple genetic variation behind it. Unfortunately, at present, we know little of the underlying genetic neurophysiological (proximal) characteristics that generate the observed intelligence scores. We have seen here that major social changes over time and in the environment produce major changes in test scores. Intelligence tests measure at least some degree of cumulative experience; their relation to genetics is as yet unproven.

One final comment on intelligence and education. The sum total of the studies reviewed here points to the importance of environment in shaping people's test scores. Such a conclusion is independent of any influence one wishes to assign to genetic factors. The importance of environment speaks directly to so-called affirmative action or preference policies in selecting, for example, students to be admitted to universities and colleges. While it is obvious that the environment shapes, or rather misshapes, the test scores and grades of minority applicants, it should be borne in mind that it also shapes the scores of members of the majority class or culture. We know that applicants who grow up in educationally favored, socioeconomically above average, and generally environmentally "prosperous" homes will score better on "objective" tests and grade averages than applicants not so favored. Thus, the abolition of affirmative action programs for minorities leaves an admission policy that is tilted toward preferences for the white middle- and upper-class populations. A truly prejudice-free admissions policy would eliminate that last vestige of preferences and would admit anybody who wishes to enter college (as was the practice until fairly recently in some European universities). Whether one wishes to advocate such a policy of open enrollment and some of its admitted problems is another question, but it does seem to offer the one truly unbiased mode of operating.[19]

Sex and Sex Differences

One of the most obvious biological characteristics of humans is their sexual identity and behavior. Therefore, the degree to which sexual behavior and characteristics can be modulated by social factors tells us something about the malleability of our biological makeup. The extent of this modification can be seen in the variety of sexual practices both within a culture and among cultures.[20] The most extreme "deviation" of what we consider to be sexual norms is found among the Dani people of West New Guinea. The Dani do not engage in frequent intercourse; in fact, they abstain from intercourse during the first two years of marriage and also for four years or more following the birth of a child. There is little or no masturbation or homosexuality. This low level of sexual activity is characterized not by repression or active suppression, but rather by an apparent lack of interest in an apparently healthy but sedentary society.[21]

I do not intend to discuss the biology of sex differences at great length, primarily because in regard to the major issues addressed here, the primary constraints on human nature, there are not any important differences between the sexes. There are no claims in the literature on either side of the nature/nurture fence about differences in consciousness, the construction of emotion, object permanence, and so on. In the area of sex differences—and of course even more so in the area of racial or ethnic differences—it is sometimes forgotten how very much alike, rather than different, all the members of our species are.

The sex differences in cognitive functioning that have been ascribed in part to our biological makeup are few. And it should be noted that these differences have been almost exclusively studied within industrialized Western societies. In other words, even the differences that have been found may, to some extent, be a product of our culture and society.

The major differences that have been found in our society is an advantage for females in verbal ability, the acquisition and use of language, and an advantage for males in spatial ability, the perception and manipulation of spatial arrays.[22] At the genetic level, there is speculation but no convincing evidence that spatial ability is linked to a recessive sex-linked gene. However, there are marked sex differences in the specialization of the two cerebral hemispheres. Lateralization, the tendency for each hemisphere to be more strongly specialized, is apparently more pronounced in men than in women. But there is good evidence that the differences in hemispheric structure and function are related to the differences in maturation between the sexes. Finally, sex hormones affect the development of some cognitive abilities, but current evidence suggests that spatial ability, for example, depends on problems of hormonal balance more than on the sheer presence or absence of female or male hormones. What makes the issue even more puzzling is the finding that spatial ability is also related to the individual's attitudes toward and self-perceptions of female and male sex roles. Again, it must be stressed that spatial ability and other possible psychological consequences of gender show a great overlap between the sexes and should not be thought of within the same category as the major (genetic) sex characteristics such as appearance, size, muscular development, and sexual organs.

The importance of looking at cultures different from industrialized Western society has been underplayed by investigators with a biological bent and is often overlooked by their antagonists. An excellent example is found in the generalizations about male and female roles and functions in the sociobiological literature. Eleanor Leacock has taken one of the more extravagant writers on the subject, David Barash, to task by comparing his evolutionary tale with the anthropological evidence.[23] In contrast to the argument that men are (and should be) sexual aggressors, women are equally able to initiate sexual contact in about half of a wide range of societies studied. Barash's claim that women are "almost universally" relegated to the nursery and that patrilocality (wives moving to their husbands' homes or tribes) is "the most common human living arrangement" are just as fanciful.[24] The converse of patrilocality (matrilocality), as well as bilocality, can be found in a variety of societies devoted to hunting-gathering or horticulture. During the 2 million years since the emergence of *Homo habilis*, 99 percent of our history has been spent in foraging and hunting-gathering societies. If we go back only to the time of emergence of *H. sapiens*, we still have spent 90 percent of that time in the hunter-gatherer mode. The patchy evidence we have for such societies—uncontaminated by Western culture—suggests that these groups were far from relegating women to the nursery. Women participated in most of the activities of the group, sharing the hunting, as well as gathering, among other duties. Children were often seen as "belonging" to the group rather than to their biological parents, and men frequently shared or sometimes were primarily responsible for the care of the young. Egalitarian relations between the sexes were probably the rule rather than the exception.

Given these rather striking differences between our society and that of hunter-gatherers, it seems reasonable to withhold judgment about genetic differences between sexes in psychological abilities. To arrive at a reasoned conclusion about differences in spatial ability, for example, cross-cultural comparisons would urgently be needed. Unfortunately, at present such investigations are too rare to be able to use them to form any kind of reasonable conclusion. On the other hand, the pervasiveness of social attitudes about the "proper" role of women and men is such that we cannot a priori conclude that they may not, from very early on in life, influence children's attitudes toward, and contact with, tasks that encourage one or the other talent or ability. We know, for example, that male and female babies are handled quite differently in most cases from birth on.[25] I move on to a description of a stable society whose gender-related structure is quite different from that of Western societies.

The Khasi: A Matrilineal Society

For the purposes of illustrating how diverse social structure may be with respect to the role of men and women and how different they may be from our own, I present a brief discussion of the Khasi people.[26] The nearly 300,000 Khasi live in northeast India, with a documented history going back to about 1500 C.E., and have maintained a matrilineal structure until the present, despite incursion of surrounding Western customs and attitudes.

The organization of the Khasi family centers around the mother, who is the family priestess. The central figures of a clan are one's mother, maternal uncle, and father, in that order. The father is not an important figure in his children's house, and when a man marries, his mother is asked to approve his marriage; when he dies, his bones are returned to his mother's house. Social organization centers around the women of the clan, and marriages are almost exclusively exogamous, those within a clan are considered sinful, and they result in loss of property and excommunication.

Property is organized in strictly matrilineal lines. Curiously, the primary system of inheritance is to the youngest daughter, the *Khadduh*. She inherits all the property in principle, though some minor bequests may go to other daughters, and property acquired by mother or father after marriage may sometimes be given to other children. The family house and land, however, all go to the Khadduh. If she is a minor, a female adult controls her property until she comes to age. Upon the Khadduh's death, the ancestral property again goes to her youngest or only daughter. If she has no children, the property apparently passes to her mother and sometimes her maternal aunts. The men of her family are primarily active in the management of the family property, though the Khadduh always maintains at least nominal authority.

The Khadduh takes care of her family, in particular her brothers and sisters, in times of adversity, and she has special responsibility for her maternal uncles. The matrilineal character of Khasi society is further illustrated by the fact that a newly married man frequently stays with his mother until after the birth of the first child. If he is childless, his property passes to his wife and his mother. A man's major authority is wielded not in his own home, but in his mother's house.

The mythology of the Khasi also illustrates the central place of women. Tradition has it that the matrilineal system originated during the migration of the Khasi. Women were appointed the role of preserving the race and retaining their culture. The mythology of the clans usually refers to an ancestral or common mother, or to a group of sisters who founded the clan.

The Khasi social organization stands in direct contrast to Western cultures. Property and religious authority revolves around the women of the clans, and the choice of the youngest daughter as the main inheritor and bearer of the family continuity speaks in particular to the untenability of standard sociobiological accounts.

If Western society has been characterized by male dominance, which in turn brought about the feminist movement to establish a more equal social organization, would one not expect the mirror image of protest to emerge among the Khasi? In fact it has. In February 1994, the *New York Times* published an article that reported on a male movement among the Khasi arguing for an end to the domination by females. The search for a patrilineal society is described as a "distant hope" which would bring about, among other things, inheritance through the male line.[27]

It should be obvious by now that it is foolhardy at best to take the social and familial organization of other animals (whether birds, ants, or chimpanzees) as models or guides toward understanding human social organization. The variability within the human species—going from strictly patriarchal/patrilineal to highly matriarchal/matrilineal structures—is incomparable to the lack of variability of such organizations within any other animal species.[28]

Some General Comments on Individual Differences

One of the arguments made both for genetic influences on the one hand and the nonpredictability of human development on the other has been that "two children reared in the same environment can turn out radically different." Yes, they can, but they usually do not. The reason for such a conjecture is that one usually looks at differences within highly stable environments (such as upper middle-class academics). Such "radical" differences are usually much smaller than the differences between environments, if one is willing to consider the full range of environments. Two children raised in the same environment generally "turn out" to be much more alike than different. For example, consider the following environments: New York black ghetto, Chinese peasant community, British aristocratic country house, Southern California yuppiedom. Do two children taken at random from the same environment turn out to be more alike than two children from different environments? You bet they do. Environment is not just noise.

Morality, Freedom, and Power

∾

Freedom has a thousand charms to show,
That slaves, howe'er contented, never know.
William Cowper

ALL PEOPLE, UNDER all conditions, live in societies that make moral judgments and take moral positions. Humans, with very few exceptions in unusual circumstances, are concerned with problems of personal freedom and the ability (freedom) to do what one wishes. The problem then becomes one between freedom and constraints, of others preventing us from acting as we desire. All of us are, if constrained, concerned with freedom, and as a result faced with issues of power and powerlessness. For whatever reasons, then, problems of moral behavior and personal freedom become part of our nature. I discuss two general issues in this chapter: the sources of human moral positions and an approach to a psychology of human freedom.

Morality and Human Nature

The issue of morality is concerned with general, often quite abstract, principles which are used, purported to be used, or advocated to be used, for the justification of specific positions on important aspects of human social life.[1] In most societies these moral principles are enshrined in the tenets of organized and unorganized religions, the dominant spiritual guideposts of the society. This is true of the Judeo-Christian tradition in the Western world, just as it is true of the Buddhist, Islamic, Hindu traditions in some other cultures. In looking at the

Western tradition in particular, can we discern a theory of human nature in the principles of the Judeo-Christian religions?

The cornerstone of our ethical traditions is embedded in the Ten Commandments. If we look at these, we find a number of very specific and apparently clear-cut prohibitions: do not kill, do not steal, do not be greedy, do not commit adultery. A little consideration suggests that behind these prohibitions is a negative view of human nature. Prohibitions are necessary only if the people addressed (all of humanity) are likely to engage in the activities that are prohibited. Otherwise, there would be no need to prohibit them. After all, there are no commandments that instruct us not to forget to eat, drink, procreate, or seek shelter. These activities are engaged in because we are biologically needful of them. But at the same time the commandments imply that we tend to be killers, thieves, and adulterers. It is a view of human nature that needs social constraints to prevent it from erupting into an anarchic, destructive social order.

In the Western tradition we have inherited a view of human nature that requires severe social constraints, that sees human beings as endowed with destructive tendencies. This is one end of the continuum of possible views of human nature. Humans can be seen as basically "bad" (destructive and competitive), "good" (cooperative, helpful), or "neutral" (subject to the particular social conditions they have inherited or face). Points of view that put us on the "good" side can be found in Buddhism and in Rousseau, among others. The "neutral" point of view is represented by the philosophical traditions of Nietzsche and Marx.

The Western position of the "bad seed" has been reinforced by various branches of Christianity and also by the more vociferous defendants of capitalist economics (but not necessarily by Adam Smith). We once again encounter St. Augustine, who was identified in chapter 2 as introducing the modern notion we maintain about the basic evil present in all humankind. But his influence was even wider than that.

As I mentioned in chapter 2, early Christian belief endorsed a generally pluralistic religion. Most of the early church fathers interpreted the Fall as the choice of Adam to opt for freedom of choice, an election for which he (and Eve) paid for by being banned from the Garden of Eden. This interpretation was a message of personal freedom to choose and was particularly powerful in the increasingly repressive Roman society. With respect to sexuality, some of the early church fathers disputed the doctrine, exemplified by Jesus and Paul, that celibacy was superior to procreation and marital relations. Jovanian called it a "novel dogma against nature." Thus, debate was fairly free. The major influence in establishing a new orthodoxy was St. Augustine, who saw in his early life of lust and debauchery the evil influences of sexual freedom and considered his conversion to Christianity in part a realization of his bondage to Adam's original sin—the rebellion against God. It is not that we are capable of, and likely to, sin, but rather that Adam's sin condemned all humankind—we are all inevitably sinners by being (biologically) human. The taint of original sin cannot be removed by conversion; it is always with us and needs to be constantly controlled and guarded against. Augustine's was an argument against personal freedom. He argued that it is "advantageous for [man] to be submissive, but disastrous for him to follow his own will, and not the will of his creator." This conception of the unfree human also served the beleaguered Roman state, with the (now) Christian emperors being able to call on religious doctrine to control and constrain its citizens.[2]

Thus, one of the consequences of Augustine's dogma was that morality is often defined by the state, and we see how state and religion can mutually reinforce their beliefs, attitudes, and powers.[3] Constitutions and laws promulgated by secular states also embody theories of human nature. They define what the society needs in order to control "bad" behavior; laws implicitly delineate likely human actions that need to be controlled, allowed, or constrained. Again, if the behaviors that the laws prohibit were not likely to occur, or if they are not thought to be destructive to society, no laws would be needed.

What evidence can be adduced from an evolutionary point of view for the emergence of moral rules and behavior? To the extent that these are rules that are part of an individual's knowledge base, we ask about the origin of these rules and principles. To what extent do we share behavior with other animals that is at least related to the moral rules we would like to see exemplified in human beings? The most ambitious (and successful) attempt to show this correlation has been presented by de Waal, who reviews the evidence on "moral" behavior by primates and then concludes by noting the following "tendencies and capacities" in other species:[4]

- Sympathy-related traits such as attachment and special treatment of the disabled
- Normative characteristics such as social rules and anticipation of punishment
- Reciprocity, as in trading and revenge
- Peacemaking exhibited in conflict avoidance and negotiations

It is reasonable to conclude that our common ancestors shared some of these moral behaviors and that we share with other primates behaviors that point to a common ground for morality. Having said that, the fact that altruistic and moral behaviors can be found among primates does not make them "moral" in the classical ethical sense, but neither does this fact make humanity moral. Just as in humans, other species who display such moral behavior also frequently deviate from moral dictates (if there are such). And it is a big step from an optimistic observation of peacemaking and reciprocity, for example, to an agreed-on moral code that is laid down in rules and laws. The development of the latter takes extensive historical and cultural background and development.

Unfortunately, the inheritance that we share with other primates does not prevent human societies from institutionalizing evil as well as good. Nothing about beliefs in moral goodness prepares us for the fact that societies can incorporate practices that seem to others totally immoral and evil. One of the more recent—and qualitatively most extreme—examples has been the Holocaust practiced upon European Jews by the German nation. Antisemitism was part of

German culture during several centuries preceding the advent of the Nazi regime, and violent antisemitism "was deeply imbedded in German cultural and political life and conversation, as well as integrated into the *moral structure of society*" (emphasis added).[5] The reassuring aspect of that example is the fact that this particular moral aberration has apparently been effectively eliminated for the majority of Germans within a couple of generations.

Rationality and Universality

I am resistant to discussing the issue of rational moral theory because there is not yet any reasonable agreement on what a rational argument or a rational theory is—from a psychological point of view. However, the simple values that I addressed in chapter 7 are sometimes taken as the basis of the emotivist argument for moral theory. MacIntyre has characterized emotivism as "the doctrine that all evaluative judgments and more specifically all moral judgments are *nothing but* expressions of preference, expressions of attitude or feeling, insofar as they are moral or evaluative in character."[6] As I have indicated in chapter 7, "evaluative judgments" may have a number of different bases, and while it may be the case that some moral judgments or principles are based on emotive grounds, they are not the simple expression of preferences. On the other hand, there is little doubt that many moral judgments have as their background impulses and preferences that are, in part, constrained by a set of moral rules. To emphasize, then, moral behavior, judgments, and rules may have a variety of different sources and origins.

My interest has been the psychological basis of preferences, feelings, and attitudes. I do not intend to go much beyond such an investigation, and in particular my goal is not to advance psychological states such as feelings or preferences as constituting moral judgments. MacIntyre criticized the emotivist position and its poverty in providing a basis for a system of moral judgments. The arguments I have advanced for the origins of values are, by themselves, insufficient to justify a system of moral judgments. Nor do I claim to be able to explicate the abstract concept of morality; instead, I am interested in understanding what behavior and thoughts are accepted as moral. But since the moral is often related to the rational, what is meant when we speak of rational judgments and values?

No clarification of rationality has significantly improved on James's definition of the core of rationality as the assertion that the world is intelligible "after the pattern of *some* ideal system."[7] As James adds, the question then is "which system?"[8] Within the present framework, a system of moral values is a system of social and evaluative schemas. These schemas are built as a function of the specific social and historical experiences of the individual as a member of a particular social group or groups. A moral value system also requires two other characteristics: cognitive consistency and conscious accessibility. To justify a moral value, we require that it be consistent with other values, and, by definition, such a moral system must be accessible for conscious constructions.

Finally, I need to address the question of universal ethical or moral values. It is the actions, thoughts, and ideas of people, usually at first in small groups and

later growing to encompass more and more of the society and culture, that define what is valued, what is a contradiction, what becomes in an important sense
familiar and accepted because it is familiar. In another language, it is the infra-
structure that creates the superstructure, and a stable superstructure harbors (or
rather implies) the reassurance of absolutes, of transcendent values. If and when
it is stabilized, such a system escapes relativism because the social values are seen
as unchallenged. I assume that the only other escape from a disordering relativism
is a stable ideology that orders actions and values under some superordinate
scheme that is socially accepted and acceptable.[9]

A cognate position has been developed by MacIntyre, who has mounted the
most recent attack on attempts to find a universally acceptable set of moral principles.[10] He argues for ethical principles that are developed and accepted by a
narrower group of people—those formed by a common "tradition." By tradition MacIntyre refers to the historical, social, and cultural background that informs a particular social grouping and its moral principles. The "traditions" described by McIntyre are organized hierarchically—by family, kin group, nation,
etc. The rules or laws of each level regulate the conflicts and differences among
moral and ethical values; these are agreed-on ground rules to avoid life-threat-
ening and other dangerous conflicts. Attempts to do that at the international level
are seen in the League of Nations and the United Nations. Such a view is, of
course, consistent with the slow and accretional development of values.
MacIntyre implies that relevant groups of people are likely to find and subscribe
to a set of moral principles that do not depend on or make a claim for a priori
universality. He decries any attempt to find moral principles that are likely to be
"found undeniable by all rational persons."

The formulation of acceptability by "rational persons" seems to describe most
universalist attempts, though Nagel considers this a travesty and prefers "to construct gradually a point of view which all reasonable persons can be asked to
share."[11] I find this formulation no more appetizing than the "travesty." There
does not really seem to be much difference between a system that is undeniable
by rational persons and one that is shared by reasonable ones. As a psychologist
I have great difficulty understanding any acceptable definition of "rational" persons, nor am I sure that definitions of "reasonable persons" are ever untainted by
the social values of the moment.[12] There is no Archimedian place "to stand"
from which we can move values or see them unshaded.

Freedom, Constraints, and Power: What Does It Mean to Feel Free?

During the past decade we have witnessed events that have called forth great
emotion, in particular, the political upheavals in East Germany and Europe, as
the constraints of foreign and domestic repression were lifted. The emotional reactions to the attainment of freedom were particularly pronounced in those
countries that had previously experienced liberal democratic freedoms, such as
Czechoslovakia (until 1938) and East Germany (until 1933). Apart from the results of wars, such as the relatively brief, though brutal, German subjugation of

most of Western Europe, these emotions were somewhat alien to our experience in modern liberal democratic states. Though not true for all citizens, constraints on political and personal liberty in liberal democracies have been relatively tempered and infrequent. The absence of constraints tends to be taken for granted.

Imposition of restrictions and repression, as well as their removal, in modern nation states presents us with a challenge to our understanding of a psychology of emotion which we have not faced before. There should be no doubt that what we witnessed, and what the subject populations experienced, were truly intense emotional experiences. I address a psychology of freedom both in the context of our understanding of the liberal values of freedom and in terms of the effect of constraints on such liberties, and the removal of those constraints, on emotional experience.

To venture into an analysis of the psychological meaning of human freedom is daunting indeed, and I do so tentatively. However, none of my discussion is intended to intrude on the use of such concepts as liberty, freedom, and power within political, sociological, or philosophical contexts. Arguments about the relative advantages of liberty or the use of power in democratic and other societies are on a different level from those advanced here.[13] Nor does this discussion address the classical free-will problem and the related notion of psychological determinism. I am concerned with the issue of personal, psychological meanings of political freedom and power.

My attempt is motivated by my commitment to understand the mechanisms and processes that shape everyday human thought and action. As I have indicated earlier, purely evolutionary or genetic approaches reveal a certain reluctance to engage in psychological analyses and theory; and one should not accept apparently inescapable (but often untestable) biological explanations until one has explored in depth other alternative explanations. I cannot let go unchallenged an appeal to an innate "instinct for freedom"[14] any more than I can accept unidimensional evolutionary stories about "human aggression," or accounts about the apparently unheralded, sudden, and full-grown emergence of human language. In general, complex human actions and thoughts are the least likely candidates for unique, singular evolutionary events. And the concept of freedom is surely characterized by complexity and multiple determination.

Natural or Constructed Liberties?

I start with a distinction between two different concepts of human liberty: a negative and a positive one discussed innovatively and at length by Isaiah Berlin,[15] though for reasons elaborated below I call them natural and constructed liberties.

The central concept of freedom in the modern Western liberal tradition was defined by John Stuart Mill when he stated that "liberty consists in doing what one desires." However, to do "what one desires" requires that no impediments be placed in the path of one's desires. As a consequence, other thinkers have pointed out that this "liberal" definition is negative, in the sense that such freedom (or liberty) requires "the absence of coercion or constraints imposed by another person."[16] Such absence may be enlarged to "the absence of obstacles to possible choices and activities."[17] But the classical reference is always to the ability to do as one wants without constraints; the crucial event is the constraint that does not happen. I primarily address this "negative" (natural) sense of freedom. It is understood, of course, that society generally and normally will accept some obvious constraints, such as the overworked constraint against yelling "fire" in a crowded theater.

Before dealing in some detail with the negative freedoms, I need to address the positive ones. Positive or constructed liberty involves the sense in which I consider myself free to act in the service of more abstract reasons, purposes, and goals, or in the interests of a higher good or higher self. The notion of these constructed freedoms can be traced to classical Greek ideas that freedom involves knowing the "good" (through learning and social experience) and deliberately choosing it. Some examples from major contributors to the Western tradition[18] will assist in defining this concept. Epictetus states that "freedom is not acquired by satisfying yourself with what you desire, but by destroying your desire"; Spinoza said that "I will call him free who is led solely by reason"; and Rousseau observed that "obedience to a law which we prescribe to ourselves is liberty." As Berlin has eloquently argued, positive or constructed freedoms may require commitments to higher ideals or goals, which themselves may be corrupted and which may be subject to abuse, both frivolous and evil.[19] Hartley Coleridge summed up the problem by defining freedom as a "universal license to be good," which leaves open to mischief or chance the public or personal definition of what constitutes the "good." More generally, "if 'freedom' becomes available for anybody's moral or political ends, then . . . all will agree that liberty is a supreme good, but they will agree on nothing else."[20] I prefer to call the negative freedoms "natural" because they arise out of the absence of constraints, and positive freedoms "constructed" because they arise out of social constructions. Natural freedom means the absence of constraints, and it is "natural" because universally it involves emotional reactions to constraints and their removal. The constructed freedoms, on the other hand, are socially constructed; they vary from time to time and from society to society; they often reflect contemporary social conventions and mores, in terms of current views of human nature, rationality, or social structure. The prohibitions involved in natural freedoms are, to be sure, also frequently social products, but the reaction to these constraints and their absence is not.

Constructed liberties are not usually the occasion for true emotional experiences; they are essentially cognitive in structure and intent. However, some constructed liberties may be constrained just as natural ones are—that is, they have both natural and constructed components, and as a result they may engender emotional reactions. For example, it has been argued that positive or constructed freedoms endow the individual with the power or ability to be part of the political process, particularly in decision making. In that sense individuals are free to the extent that they are in a position to participate in government, in the distribution of public goods, and importantly in necessary restrictions of natural liberty. And at least some of these liberties may be constrained. However, these constructed liberties may refer more to the empowerment of individuals in the political process rather than to their "wish" to participate in it.[21]

I am concerned here primarily with a psychological understanding of the "negative" or natural liberal sense of freedom or liberty.[22] I should note, though, that this does not imply that I accept this definition as the only or the preferred sense. However, the concern with this particular use in the Western tradition motivates my attempt to understand its psychological underpinnings. I return later to possible alternative social goals.

The Psychology of Constraint and Freedom

To be constrained from initiating some action or from continuing to act is a prime example of discrepancy and interruption, and I approach the emotional consequences from the point of view presented in chapter 6. One usually expects to initiate an intended action and to continue and complete an action once it has been initiated. It is irrelevant to the present discussion whether one wants to act or whether the action is automatic or deliberate—questions of voluntarism or free will do not affect the fact that an action is initiated and prevented from completion. We wish to be able to do "what we want." With respect to freedom or liberty, constraints are actions or conditions that prevent the execution of a particular act or a class of behaviors. Prohibiting people from associating with one another, keeping them from traveling abroad, confining them because of their unpopular or prohibited beliefs, restricting the occupations in which they may engage, are just a few examples of constraints that denote the absence of freedom. All of these constraints are discrepancies or interruptions of hopes and expectations, actions, and incipient actions.

The emotional consequences of an absence of liberty (i.e., the presence of constraints) are obvious. Actions are contemplated or even tentatively initiated and are then interrupted or diverted. As a result, arousal occurs, and the ensuing emotional reaction is negative. The negative affect is, of course, a result of the negative evaluation of being prevented from doing what one wants to do. The most frequent emotional reactions are resentment, frustration (the sense of being blocked), a sense of oppression, and, very frequently and partly related to all of them, anger (the sense of wishing to remove or destroy the blocking agent). The emotion of anger is usually associated with being blocked from reaching a desired goal, and the central value is presumably the desirability of the goal—the freedom to do what one desires to do. To the extent that we fail to have any alternative action available, the very consideration of a prohibited act generates anxiety; we imagine the desired state of affairs, but we also become aware of the obstacles that prevent the achievement of this state and of our impotence in the face of these obstacles. Such a view restates part of Kierkegaard's discussion of freedom. Kierkegaard related freedom to our ability to become aware of possibility (i.e., possible choices).[23] We create possibility, and the very consideration of such possibility generates conditions of anxiety and dread.

It is a truism that living in a human society means living with constraints. Common cultural norms and social roles constrain a vast array of behaviors, beliefs, and attitudes. All societies constrain. They prescribe pervasive and commonplace concerns: how we act as a function of our gender, what we wear when and where, what are acceptable foods, how and when we vote (if we do), whether we live in apartments, tents, or houses, how we form families, whom we can or cannot marry, and how we engage in social and sexual intercourse. Many of these prescriptions produce no conscious constraints and are not discrepant; they are accepted as "right and proper." In the areas of social and political thought and behavior, we encounter the same dialectic of the familiar and the discrepant that I discussed in chapter 7 with respect to individual behavior. There are aspects of society and culture that are accepted unconsciously without deliberation or further consideration. These aspects (including some areas of food, dress, and social intercourse) provide the acceptable, the "familiar," and even the reassuring features of our daily existence. The question of whether these cultural constraints are desirable or not does not, in the majority of cases, even arise.

It is possible to roughly divide the various constraints that exist in society into three categories. First are the cultural norms just discussed; they do not give rise to discrepancies—there are typically no wishes or desires that are discrepant with them. Second, there is a set of constraints on choice that we encounter that are either not of great importance—we care, but not too deeply, about being constrained in these desires or wishes—or that are constraints of only temporary, although possibly great, importance. We may be conscious of these constraints, which include some social rules (do not cross when the light is red), as well as idiosyncratic obstacles. Among these are the many daily annoyances in which we are prevented from doing what we "wish to do." These too vary over a wide range and include such desires as wanting to exceed the speed limit, wanting—as a child—to stay out after midnight or to eat only chocolates for dinner, wanting to let one's hair grow long, wanting to smoke in a restaurant, and so on. Finally, there are the natural liberties, essentially the political freedoms. These are lasting, important, and considered to be part of our "rights" as citizens in a democratic society. The constraints imposed on the first category are rarely considered as such, those in the intermediate category refer primarily to restrictions on group or individual preferences and wishes, and those in the third category are generally considered violations of "rights."

The "habits" of the first category outlined above are rarely subject to the kind of emotional reactions that I have sketched here, but the wishes and rights of the second and third categories usually evoke emotional experiences. When these wishes and rights are constrained, we are faced with discrepancies, autonomic arousal, and subsequent emotional reactions. The emotions may vary from embarrassment when we exercise our social norms in a different social or cultural setting and discover that they are not acceptable, to negative feelings of a usually temporary nature when our desires are thwarted, to lasting social disaffections when rights are blocked or frustrated.

It is also the case that the habits, desires, and rights are subject to change that may be slow or may take place rather suddenly. Cultural norms change over time and personal desires change—for example, as we shift our position in the social group. But political rights also change as societies and their structure changes.

Political rights were different in Roman society from medieval times and different from modern rights, and we should keep in mind that in terms of recorded history the liberal concept of freedom is relatively recent. Furthermore, some of the apparently nonpolitical desires (and their constraints) may become politicized, as, for example, length of hair became both a political statement and subject to social constraints in some corners during the 1960s.

To return to a more specific question about the psychology of freedom, what are the emotional conditions when natural liberty does in fact exist? Negative liberty implies the ability to complete actions that may or may not have been frustrated at some earlier time or to engage in actions that are now and always have been available. I want briefly to consider the latter—liberties that are in fact available and have rarely if ever been constrained in the past. These are not usually considered as liberties—they are taken for granted; they actually become part of the cultural background and are accepted unconsciously. In addition to this usual cultural background, in most established democracies the choice of (available) foods, the selection of reading materials, and similar everyday "liberties" produce little if any recognition as freedoms, and no affective reaction. Societies that have little history of previous constraints or authoritarianism should show little emotional or affective reactions to the liberties available in everyday life, just as there is little reaction to the other "habits" and norms of the culture.[24]

Now consider the cases in which liberties are available that have been previously constrained. Such events may occur in nearly all societies. The lifting of rationing restrictions after a war permits people to do things they were previously constrained from doing. The emotional scenario may be quite complicated. On the one hand, prior experience leads one to expect constraints which are now removed. This is just another case of the violation of an expectation, with the attendant arousal, but the evaluation and the ensuing emotional state would be positive. On the other hand, the very ability to carry out an act that had previously not been possible produces the evaluative cognition of completion[25] which, together with the arousal derived from its relative unexpectedness, generates a "joy of completion." In the case of societies that switch—slowly or suddenly—from repression to freedom, these effects are much greater; the emotional reaction is much stronger (if for no other reason than that the arousal will be much greater). The best recent example is, of course, found in the European revolutions of 1989. This sense of freedom, expressed in Martin Luther King's "free at last," of being able to act as one "wishes," produces the positive sense of freedom that natural liberties can engender.[26] It should be clear now that the "freedom emotion" is more likely to occur under conditions where previous constraints and prohibitions have existed. The values that inform this emotional reaction include the desired end goals (e.g., unconstrained speech and association) as well as the more general removal of constraints in principle.

There is another side to the psychological reaction to a lack of freedoms. I have noted in chapters 4 and 7 that extensive experience with situations and events generates very stable schemas. And stable schemas generate the stable aspects of society—a condition necessary for the very fact of social organization. Successive encounters with such events will produce an easy "fit" with the underlying schemas, and will generate the subjective experience of familiarity—and eventual acceptance. Briefly entering the arena of sociological and political analysis, I offer the following only as illustrative and by no means complete.

Authoritarian societies that generate basic satisfactions for their members will, despite many constraints and prohibitions, produce acceptance of the status quo. This may be particularly true if the government in power can produce satisfaction of other (real or imagined) desires or wishes. The widespread acceptance of a fascist regime in Germany during the late 1930s existed in the absence of many natural freedoms. Their absence was counterbalanced by the availability of employment, the appeals to German hegemony, the defeat of "foreign" influences and domination, and promises of increasing power to the German nation.[27] The lure of familiarity, coupled with obvious (but often superficial) benefits, may lead to the acceptance of a state of affairs which, objectively, constrains many actions and restricts freedom. In addition, the restriction to certain permitted actions, the satisfaction of perceived needs and desires, and the power derived from the protection of a powerful government will be perceived as producing other benefits. If one is at ease, to some extent, with current life, one need not entertain novel, unfamiliar actions with uncertain outcomes—which may themselves lead to arousal and failure. Stories of long-term prisoners reluctant to leave the familiarity and "safety" of prison illustrate how constraints may be preferred to the uncertainties and dangers of the "free" life.

There is a sense in which the citizen of a fascist state and the prisoner in jail are psychologically "free." They do not, for different reasons, request freedom of speech or of movement, and as a result do not experience the deleterious effects of the constraints that are in force. The constraints have become cultural norms. Such an analysis does not, of course, address the political or philosophical issues involved in the absence of these freedoms.

The Psychological Meaning of Power

The exercise of liberty—the absence of constraints—requires controls over the means of attaining desired ends and goals. In its most general sense, "power is the ability to pursue and attain goals through mastery of one's environment."[28] In addition, power may be defined as the possession of desired goods and the control of scarce means and of the conditions for action and communication.[29] But "[w]hen there is conflict between individuals and groups for possession or control of scarce means and conditions of action, control over means is a condition of the availability of alternatives, and hence of choice and freedom."[30]

Groups that have "power" and privileged access to the goods of society protect that access by the domination and restriction of less powerful groups. Under those conditions, it is the groups that hold power who enjoy maximum freedom in a society. Conversely, the powerless will be deprived of goods and constrained from exercising actions to the extent that such access and exercise restricts the "freedoms" of the powerful. In Berlin's felicitous phrase: "Freedom for the wolves has often meant death to the sheep."[31] But, as I have noted above, in a stable society the powerless may accept the constraints and are, in that sense, psychologically free. It is only if one posits some basic human needs for expression, sustenance, shelter, association, or movement that the associated freedoms can be said to be generally relevant to people under all conditions.

The psychological satisfaction for those in power is the exercise of freedoms, particularly when such "freedom" becomes evident by the fact that others in the society are constrained from its exercise. And it is irrelevant whether these freedoms are trivial or crucial. It is the mark of subjective satisfaction by the powerful in a restrictive society to enjoy the use of a private car as much as the availability of adequate food, the privilege of criticizing "their" government as much as the availability of sumptuous houses. In the extreme cases, feudal rulers were often able to command any satisfactions available in their societies. This is not to say that the problems and contradictions of power do not generate their own conflicts and dissatisfactions, but the distinction between powerful and powerless groups is obvious.

Since the last century, the primary analysis of power and control has been in terms of economic classes. Marx saw the basic locus of power in the control over the "economic surplus" through the control of the means of production. Subsequent thinkers in that tradition have tended to subsume other power relations under the economic ones. However, it has become increasingly clear that, whereas economic dominance is important, dimensions other than control over the economy have to be considered seriously. Control and power have been exercised by whites against blacks, men against women, hetero- against homosexuals. In all of these cases, privileges and controls (including, of course, economic ones) are preserved by the powerful over the less powerful groups. The powerless experience less freedom because they are constrained from exercising certain roles and actions, and therefore they tend to experience negative states.

If power is identified as the ability to restrict others and to entertain actions without restraint, then membership in groups of power is something that is likely to be desired by those who are constrained. More generally, though, it appears that actual membership is not necessary to derive some (at least temporary) satisfaction from identifying with powerful groups.[32] If I feel constrained and deprived of goods and the possibility of actions, then I may well seek some other role or social identity that will restore at least a sense of possible power. Such identifications are more likely the more confused the state of society, the more uncertain one's future status. We see such identification with actual or potentially powerful groups in the conflicts that have swept the world in recent decades. Catholics versus Protestants in Northern Ireland, Serbs versus Croats versus Muslims in former Yugoslavia, Azeris versus Armenians in the former Soviet Union, Hindus versus Muslims in India are just some examples of where identification with groups are seen to generate power, which implies empowerment to followers of those groups. Class, nationality, religion, and other groupings represent social forces with varying salience and attraction at different points in history.[33] Members of such groups, in general, are likely to find a sense of empowerment in group membership, and such group membership may in fact give them "liberties" (lack of constraints) they would otherwise not possess. Conversely, the powerful ruling groups may confer partial "rights" to the powerless, in the expectation that even a minimum sense of being freed of constraints will divert attention from restraints actually still in force.[34]

Alternatives and Additions to the Concept of Liberty

The two major "democratic" revolutions of the eighteenth century nailed the cry for natural liberty to their banner.[35] As part of both "liberty, equality, and fraternity" and "life, liberty, and the pursuit of happiness" they ensured the (at least partial) triumph of the liberal concept of freedom.[36] But, as we have seen, that concept speaks to rights that are defined by the absence of complementary constraints. And it appears to have overshadowed in the popular imagination the other goals of those powerful movements, such as fraternity and the right to "happiness." Are these human goals and rights that can supplement natural liberties?

I have already indicated the problems with constructed liberties—the sometimes pernicious ends that may be traps of higher goals. On the other hand, constructed liberties do speak to informed principles of human association and possibly to rights of participation in decision making and to access to public goods. The political question—not to be addressed here—is to define such higher goals and to assure such rights and access without falling into the trap of subverting the public good in the name of some higher good. Political and philosophical anarchism and also its contemporary offspring, the often unfraternal and antisocial "libertarianism," see their goals in the free association of people and the absence of concentrations of power in groups or state machinery. A similar goal—the classless society with the state having "withered away"—is represented in Marx and his followers. But we have experienced how such a goal can be subverted into authoritarianism in the service of final or higher goals.

When constructed liberty is defined as "knowing the good," such knowledge implies deep understanding, educated choices, and social consensus. With these conditions, the subversion of constructed liberties might be avoided. The analysis of such "knowledge" is a complex cognitive problem, requiring an analysis of how we understand the sociopolitical and economic world and how we develop our sociopolitical consciousness. In part, the recognition of group rights—the empowerment of significant social groups—might lead to an understanding of group freedoms. Such conditions and extensions might well be applied to constructed liberties such as fraternity, equality, and the pursuit of happiness. They embody ideals that exclude constraints and that envisage access to goods and participation in decision making. Their psychological consequences have rarely been pursued, but they are likely to include a variety of social satisfactions. Fraternity responds to the social nature of human beings, who are biologically and psychologically bound to the social matrix. Its exercise may generate such emotional satisfactions as helping others, being a support to one's neighbors, avoiding actions that may do physical or psychological harm to others, and the positive values of contributing to and benefiting from cooperative social action.

One aim of this chapter was to show how the analytic, constructivist approach can be useful to our understanding of complex individual and social phenomena. In the process, we have seen how social constraints and proscriptions can act as powerful discrepancies and interruptions. The exploration has also led to a further discussion of values such as the negatively valued presence of constraints on the one hand and their positively valued absence on the other. This chapter should have made clear why I consider the avoidance of constraints and the desire for freedom basic to human nature in general. This is not because of any innate instinct for freedom but rather because constraints that prevent people from acting as they choose produce negatively valenced emotions. Avoiding these aversive emotions also generates a desire for personal freedom. Thus, our values usually include the ones embodied in our views of freedom. And I have tried to show how these values operate in individual and social thought, action, and emotion.

Human nature as exemplified in the desire for political freedom illustrates again the intimate intertwining of nature and nurture. Cultural constraints and our reactions to them are exaptations derived from social and evolutionary history.

ELEVEN

Cognition and Language

⌒

Language is the dress of thought.
Samuel Johnson

THROUGHOUT THIS BOOK my intent has been to discuss primarily those aspects of human thought and action that are likely to be culture free—that is, humanity wherever it lives, loves, works, and creates its social fabric. And whereas more work has probably been done in the past 40 years in American psychology on so-called cognitive functions, most of them are very much part of the context of Western society. This is true of much of the work on memory, problem solving, reasoning, and even language acquisition and comprehension, and the study of values in Western psychology presented in the Appendix illustrates the roots of this cultural dependency.

Consider the topic of memory, for example. The field of research that covers memory from amnesia to story comprehensions is still in a state of flux. Little of consequence had been achieved on the nature of memory under the aegis of the behaviorist dicta, and after the rush of research in the 1950s, 1960s, and 1970's had quieted down, still no consensus had been reached, though some potential areas of positive knowledge emerged. Academic research, which used to be preoccupied with the memorial recovery of lists, has begun to realize that much of human memory is not deliberate and is concerned with organized mental contents rather than with the deliberate recovery of discrete words and items. It can also be argued that the realm of dreams belongs to this rubric of recovering stored memories.[1] Other cognitive and mental phenomena are too obscurely known, though widely practiced, to be the subject of any specific assignment to current assertions or to evolutionary speculations[2]—though it is surely the case that such cognitive activities as memory, language acquisition, and the assignment of causality are among universal characteristics of the human species and have large evolutionary and genetic components.

Mental mechanisms are difficult to assign to specific evolutionary processes. Obviously a brain/mind that can perform certain functions is the result of both evolutionary and environmental pressures. In order to make a strong statement about their evolutionary or genetic basis, however, one needs to show strong universality across radically different cultures. I have noted before that universality alone does not imply a genetic basis, but without universality it is difficult to argue for genetics. At the same time, one needs to be careful that a possible universal mental trait is not obscured by some culture-specific development. Thus, it is clearly not easy to make genetic claims for mental mechanisms, and surely one cannot do so without extensive cross-cultural evidence. The latter is usually missing from claims about the genetic basis of mental functions. In what follows I try to confine myself to some, but by no means all, human cognitive functions that are likely to be part of human activity in most cultures and societies.

Categories of Thought

With these reservations in mind, I discuss here the three ways in which the human mind organizes the world around it. Because it is unlikely that any other animal similarly categorizes information, I include categories of thought here as relevant to the notion of human nature. The organizational structures that are used to store information about the world and that are used in memorial retrieval can be grouped into three general prototypes:[3]

1. Subordinate structures organize information hierarchically. Particular events are located as subordinate or superordinate to other parts of the structure. Subordination is probably the most widely used organization in Western science; we use it to order our knowledge of plants and animals, family structures, and even administrative organizations. Subordinate structures are typically represented in the organization of common categories, but are also assigned to the structure of plans and goals. These hierarchical structures were of particular interest in the psychology of the 1960s and in the "early" days of uncovering the structure of mental representations.

2. Coordinate structures represent mutual or symmetric relations among relevant events. These knowledge representations are unitary, holistic organizations in which each constituent is a necessary part of the structure. Visual representations are typical of such structures, as, for example, our knowledge of the layout of our house or garden, or of a friend's face. The relations among all the constituent elements define the organization and are coordinated in their use and effect. No part of the structure is in itself sufficient as a mental representation, and all, or most, parts are necessary for a complete representation.

3. Proordinate structures are represented by serial organizations. The representations are linear, and much of language, from the alphabet to syntactic struc-

tures, is either proordinately, or serially, organized or contains substructures of this type. Retrieval of any element from such an organization often requires the retrieval of a temporally or spatially prior or subsequent event.

Having listed these three prototypes, we should keep in mind that pure instances of any of them are seldom found. Rather, our knowledge and memories are typically organized along more than one of these structures. Consider the classical shopping list, which has both serial (proordinate) and categorical (subordinate) components. We list our needs by types (e.g., meat, dairy, bread), and within types may follow the serial structure of a recipe (e.g., for a meatloaf) to list specific ingredients that are needed. Mnemonic devices, such as the Roman method of loci or mnemonic rhymes, often incorporate all three of these structures. And stories, scripts, and scenes frequently are intricate combinations of all three.[4] Even structures that seem obviously serial, such as the alphabet, are represented in both hierarchical (chunked) and serial structures.

Whatever structure is used in the process of registering events or objects for later construction and use, the notion of relatedness is central to the contemporary treatment of memory. The aforementioned types of structures all have in common the function of principled relations among mental contents. These relations are not neutral links among events but rather define the kind of relationship that is employed.

Language

It is generally taken for granted that the characteristic that most readily distinguishes humans from other animals is language. Language has not appeared in any other species, even though most other animals do have more or less sophisticated communication modes. Human language is currently the most efficient way for organisms to communicate. It has the advantage over preverbal sign and body languages in that it is verbal or oral for most people, which means it requires no elaborate orienting behavior (we can hear without looking at the sender), and its specific forms are easily adapted to different cultures and societies. And while there may be similarities in underlying grammar and phonology, the relationship between language and the things to which it refers is, in principle, arbitrary, resulting in thousands of different languages.

Grammar and phonetics may well be constrained by our evolutionary history, but other aspects of language such as the details of semantics and pragmatics are probably culture specific. These latter characteristics make it possible for language to develop culture-specific shorthand expressions and references, to refer to itself, to absent objects and events, and to future events. The intricate relationship between language and culture also emphasizes the importance of language in the transmission of culture and cultural artifacts, making possible the preservation and transmission of both innovations and cultural habits. Wheels can be described and need not be reinvented. Such cultural transmission has been in the oral mode for most of the time since language was invented. Written language is very young in the history of our species, having been around for only some 6,000 years.

Evolution of Language

It is likely that as human cognitive functions became more complex, the early communicative devices such as gestures, noises, and facial displays became inadequate to express the range of categorizations and relationships. With the parallel development of a more competent vocal apparatus, human language developed. But even such speculative statements need better grounding, and nothing more definite can be said about the evolution of language at this time. Despite authoritative statements, the plethora of competing conjectures is truly astounding.[5] The speculations range from the notion of selectionist, adaptive evolution of human language by design,[6] to an innate language faculty that is somewhat mysteriously the result of an exaptive rather than adaptive selectionism,[7] to those who accept neither the innate faculty nor a selectionist account of the resulting human language, but rather adopt an exaptationist account of selectionist parts coming together in a new synthesis.[8]

Most authorities agree that language evolved in a social environment that provided pressures for better communication, that language provides an avenue to shared mental and conceptual activities, that it probably was preceded by a sign and gesture language (see also the discussion of facial signals in chapter 6), and that without it human nature would be radically different from what it is today, but few agree on how this has all come about. One possibility is the development of an expressive language of facial and body movement communicating important messages (as in some other animals) that develops into a conglomerate of semantic, verbal messages together with a rudimentary grammar of pointing and other connective body movements. These "primitive" languages (e.g., in *H. habilis*) could then develop through selective pressures and the use of other cognitive and behavioral skills into modern language. It is highly likely that *H. sapiens* and even their predecessors developed some complex communicative system prior to the emergence of modern verbal language.

The hypothesis about an ancient gestural language has a long and honorable history. Wilhelm Wundt, in the volume on Language, part of his monumental *Völkerpsychologie* (ethnopsychology), notes that "language presumably developed out of the simpler forms of expressive movements."[9] And in his introductory text, Wundt obseres that deaf and mute children, raised without any deliberate instruction, will communicate by means of "a natural development of gestural speech, which combines meaningful expressive movements."[10] A combination of miming and pantomiming signs produces both pointing and painting gestures which "generate a type of sentence construction whereby objects are described and events told."[11] Wundt spends a chapter on gestural language, its complexity and its likely place as a forerunner to verbal language, with its own grammar.[12] And in the contemporary literature similar suggestions appear, as in the conclusion that "even a limited combination of vowels and consonants, reinforcing facial expressions and manual signs, must have offered [*Homo habilis*] practical advantages."[13]

Givòn presented another approach that takes into account both ontogenetic and phylogenetic considerations of the development of grammar and puts primary emphasis on the necessary precursors of verbal grammar in the race and in the child.[14] It is unfortunate that Steven Pinker's recent authoritative exposition of the neo-Chomskian point of view is innocent of any knowledge of Wundt's and other's considerations of language origins and thus glosses over the general question of the prehistorical precursors of spoken, verbal language.[15]

Gould (in agreement with Chomsky) also assigns exaptive origins to the human language function, locating the "origin [of language] as a side consequence of the brain's enhanced capacity."[16] That position implies a single innate language device, as advocated by Chomsky. It is, however, probably the case that the various aspects of language, its physical requirement in the human larynx, its cognitive and conceptual bases, and its grammatical structure have different adaptive and exaptive bases in their evolutionary history, but even such a statement must be hedged in the light of the extensive arguments and ignorance that pervade the field.

In chapter 3 I provided some of the physical evidence for the precursors of human language. One line of thought implies that the necessary tools for language development may have been in place 1 million years ago and longer, followed by the specific emergence of language about 100,000 years ago.[17] It is highly likely that more definitive hypotheses and some possible agreement about the evolution of language will require more knowledge about the neurophysiology and structure of language.[18] What of the hypothesis concerning a universal grammar?[19] Given the research reviewed thus far, do we need to account for similarities in grammars by ascribing them to some genetic universal grammar? Instead, such similarities may be ascribed to the principle of similar solutions to similar problems. Thus, just as placental and marsupial mammals have produced independently some rather remarkably similar species, and just as such unrelated swimmers like sharks, dolphins, and ichthyosaurs have developed fins that stabilize their aquatic activities (see chapter 3), it may be that different languages in different parts of the world under different conditions may still produce similar solutions to the same problem—namely, how to organize language.

World and Mind: Structure and Representation

Both the physical world and the mind are structured. In dealing with our surroundings, the mental system represents the structures of the world, but not in a one-to-one correspondence. This distinction is important to an understanding of the way we act and think in the world and also for the way we build mental models of that world.[20]

The importance of this distinction is best illustrated if we reconsider, from a new point of view, the distinction between competence and performance that was popular among some psycholinguists a couple of decades ago. The distinction had its origin in the work of the French structuralists, who introduced the modern concern with the structure of human functions and artifacts. In turn, the structure of language, the basic language code, was taken to speak to the language competence of the individual, which reflects the structure of language as such, independent of the experience of the individual. Performance, in contrast, was seen as the actual language usage of the individual, limited by experience and by cognitive and psychological factors. At times, competence was related to language comprehension, whereas performance was seen to reflect language production. Whereas relatively little is heard of this distinction nowadays, the point of interest concerns the nature of the representations that distinguish competence and performance.

According to one interpretation, competence refers to the underlying structure of a particular ability, skill, or characteristic, whereas performance indexes the actual instantiations and realization of that structure in human thought and action. Psychologists may have been particularly intrigued by the distinction, in part because it has some analogic similarity to the learning/performance distinction that was so artfully used by the learning theorists of the 1940s. The primary distinction was between observable performance and data and inferred, internal representations of what was learned. What an organism had learned and what was actually observed in behavior depended on particular situational and individual factors at the time of performance.

The competence/performance distinction is frequently a distinction between a formalist description of a system in the language of one discipline (philosophy or linguistics) and the imputed "use" of that system by a psychological entity—the human mind. In other words, competence tends to be described in terms of the structure of the system (such as language), and not in terms of the mental representation and actual use of that system. There is increasing evidence that the gap between the domain of a rationalist theory of language and the actual performance of the human actor is widening. It may therefore be more useful to reject that distinction and to ask for approaches that are couched in psychological terms, dependent on characteristics of the human mind for the parameters of what the language users can potentially do and what they actually do.

Competence theories are likely to be perfectly valid descriptions of systems within another domain. Thus, the structure of language or logic is one domain; whether and how that structure is perceived and incorporated into the human mental apparatus (if it is at all) is an entirely different domain. A similar distinction has been made in the past between a territory and the map of that territory. Psychologists may get some hint from theories of language of the problems faced by a language or logic user, but the structure of those theories does not define the competence of the human user. A person may use some or none or all of the structure, but only empirical evidence and creative theory can make clear what is used and what is not.

The distinction between the structure of a domain and the structure that is mentally represented has been made for the case of story grammars and the mental models of these grammars that people use in representing, telling, and remembering stories.[21] At a simpler level, consider the structure of a house as it is represented in architectural plans. These plans could be used to find one's way around the house, but they are not isomorphic with, not the same as, the structure of that house as it is represented in the mind of one of its inhabitants. The structure of the house as mentally represented is what we speak of when we refer to its mental representation.[22] The architectural structure of the house is of interest because it will reveal certain parallels and divergences from the mental structure, but it does not reveal the underlying competence of the house user. Neither is the structure of a language, qua language, a description of the actual or potential competence of its speakers.

A similar distinction needs to be applied to our understanding and use of social structures. The structure of a society, of a social group, or of a culture affects the way its member behave and think; it constrains what is doable and thinkable. For example, the structure of a society determines who does what and with what tools; it determines social mobility and the relations among its races, classes, and other subgroups. On the other hand, there is the structure of that society as it is perceived by its members. Not only is this perception different among different groups and individuals, but it is only in part determined by the structure of the society. This perception (or representation) of the structure of a society among its members in turn has effects which can be distinguished from the direct effects of the societal structure per se.

Finally, I want to defend the psychological view inherent in most of my discussions and to reject, in principle, appeals to a complete reduction of psychological explanation to neurological ones. This continues the arguments initiated in chapter 4 on the relationship between mind and brain.

Reductionism: Mind Is not Brain, or Vice Versa

I return once more to the fascinating mind–body problem. The increasing interest in brain–behavior relationships and the rise of neuropsychology have both contributed to a revival of reductionism—the view that complex (mental) phenomena can be exhaustively understood (reduced) to more basic (physiological) ones.

To review the arguments in chapter 5: The reductionist argument cannot be left in isolation from other reductionist claims in the sciences. Differences between mental and physical systems cannot be understood any more easily than the reduction in any other two domains, such as chemistry and physics, sociology and psychology, or physiology and physics (to raise an extreme example).[23] A similar argument was made by Alfred North Whitehead when he noted that electrons within living bodies behave as they do as a function of the plan (structure) of the body and not because of any inherent characteristic. As I have noted in the discussion of the physical basis of consciousness in chapter 5, the arrangements of atoms in molecules are not "inherent" in the atoms themselves, and even such complex phenomena as the "handedness" of quartz crystals are the result of a chemical process; they are not "embryonically present in the atoms."[24] Qualitative changes occur as a result of quantitative ones (as in the magnetic property of iron as it is cooled). Complex organisms and complex organs (such as the brain) can produce qualitatively different characteristics (such as consciousness) that are inherent neither in less complex organizations (e.g., in some other animals) nor in their constituents (neurons).

In another related example, Putnam in discussing minds and bodies (or matter and soul), notes how ridiculous it would be to "explain" why a cube of a certain size passes through a square but not a round hole in terms of the atomic structures of the board and the cube. An explanation in terms of the relevant geometric relationships—that is, at a "higher" level—has more generality and, in fact, an explanation at the "lower" atomic level would conceal the geometric laws. Similarly, explanations at the psychological level address a different set of laws than do any possible physiological explanations of the same phenomena. Furthermore, Putnam notes that the mind is a system with so many degrees of freedom (i.e., it is so complex) that it can "imitate to within the accuracy relevant to psychological theory any structure one can hope to describe."[25]

The point to be made is obvious. Explanations at the psychological level cannot, in principle, be reduced to explanations of the same phenomena at the physiological or hormonal level. This is not to say that the cooperation between psychology and neurophysiology is unnecessary or fruitless. On the contrary, this cooperation has created and invigorated a newly rediscovered field of investigation, and observations and hypotheses have usefully gone in both directions. But, in general, it seems to be the case that "explanations" in science in terms of more "basic" processes have occurred after the more complex observations and theories were well established. This has happened in biology where gene theory was a necessary precursor of molecular explanations, in physics where an understanding of atomic and molecular functions preceded nuclear physics in the modern sense, and in psychology in both vision and acoustics where the behavioral insights preceded the physiological ones. At present, the emergence of the cognitive neurosciences followed the exploration of language and memory in psychology and linguistics.

Concluding Thoughts

My explorations of human nature have concentrated on some areas in which I have special interests and some that need to be discussed because of their pervasive relevance to the topic. In the course of putting this book together, I have necessarily neglected important aspects of human nature—and even some that I probably have not thought about at all. But the major purpose has been, from the beginning, to show how a psychological approach makes some aspects of humanity comprehensible and that an unquestioning invocation of innateness is not necessarily the best way to understand who we are and why.

It is unfortunate that much of valuable thought about humans and their nature tends to be vitiated by a lack of two kinds of scholarship: historical development and precedence on the one hand, and cross-cultural awareness on the other. Pinker's tour de force on the evolution and structure of language demonstrates both.[26] In his treatment of language, he essentially discounts any possibility of complex languagelike precursors of verbal language, despite the fact that a complex gestural language has been discussed as a verbal language predecessor since the nineteenth century. It impoverishes our view of language to see humans as devoid of sophisticated communication devices before the "miracle" of spoken language occurred.

At the end of his book, Pinker illustrates one strand of current fashion by postulating a variety of different innate modules, each of which is responsible for some complex aspect of human functioning. Postulating single modules for each distinguishable bit of behavior and rejecting general principles of mental functioning have a distinctly postmodern flavor. In any case, one should not postulate the contributions of some simple genetic formulas to account for such complex behaviors as maternal care, patriarchal predominance, sexual behavior, or phobias, for example, while remaining innocent of any knowledge of societies where children are brought up by indistinguishable multiparental units (where there is no "mother" as such), the existence of matriarchal societies, the dramatic variation in sexual behavior, and uniqueness of some Western phobias. Disentangling nature and nurture is one of most important tasks facing the biological, psychological, and social sciences; the hard work necessary cannot be preempted by facile pronouncements.

It should be obvious by now that the embeddedness of human nature in our surrounding culture makes it difficult to come to any final conclusions about the nature of humanity. Not only does our society and culture shape our behavior and, in particular, those constant and trans-situational cultural habits that come to seem "natural," but at the same time, the culture defines the kinds of questions we are able to ask about our natures. I illustrate how psychology has been a reflection of human culture in the Appendix which follows.

Psychology as a Reflection of Cultural Values

An Essay in the Social and Historical Bases of Modern Psychology

∞

Prologue

In the main part of this book I have asserted that many human characteristics are a function of the culture in which they are imbedded and acted out. The same argument must, of course, be made with respect to the hypotheses, theories, and conjectures we generate about human nature. In particular, the social sciences, including psychology, must be seen as part and parcel of their social setting. I have added this appendix to illustrate exactly this kind of relationship. Thus, I come full circle by claiming that even our explanations of human nature are a function of the historical and contemporary setting in which they are advanced.

Introduction

I intend to illustrate some of the ways in which psychology is situated in the fabric of the surrounding society and to discuss some of the prejudices and traditions that have created the present situation in psychology.[1] I discuss aspects of the history of psychology to show how various psychologies reflect the characteristics and values of the society in which they are embedded. To illustrate this theme I use three case histories: the German context of the nineteenth century that gave rise to Wilhelm Wundt and his psychologies; an examination of the social climate at the fin de siècle in America and Europe that gave rise to American behaviorism and Gestalt psychology; and the "cognitive" emphases of the last 50 years. This essay is not intended to be a history of psychology. Rather, I use historical incidents to illustrate some general points. The final section of this essay addresses the way psychology is done today; it suggests alternative ways of approaching psychological phenomena.

I am neither competent nor prepared to answer the "big question": What is it about the structure of a society—its development, technology, economy, social relations, etc.—that favors one or the other set of values? I can only try to begin asking similar questions about psychology in particular. Specifically, what are the social forces that are likely to have influenced the course of psychological science and the values that inform it? Here I present some sketches on the influence of larger social forces on the psychological enterprise. In the process, I argue that not only is psychology not value-free, but the values that motivate psychologists can be found in the society and culture that surround them and of which they are a part. And it is not just the general culture that shapes psychology, but specific influences from other sciences also play their part. The desire of psychology to be experimental is, in part, influenced by the intention to be a hard, natural science. And specifically in the recent past, the emergence of the information sciences and the development of the computer have influenced psychological thinking. Conversely, it is not just a one-way street of culture shaping sciences, but the changes in the sciences and their institutions affect the social fabric of society.[2]

I do not intend to imply that the social situation in which particular psychological points of view and attitudes developed were the causes of these views and attitudes. I believe the situation is much more complicated than that. The general themes of the culture are to be found in the contemporary social sciences, not in a one-to-one correspondence in time, but rather in the sense that similar themes will be distributed over a particular time period. In brief, I want to treat social science as much as part of the contemporary culture as its clothes, rituals, foods, music, art, ethnic prejudices, and so forth. Social science is, like many other social activities, a symptom of the embedding culture and society. The causal forces that shape that culture and society (and its symptoms) are numerous and multi-determined and far beyond the speculations of this essay. I also believe that the human (social) sciences are more subject to the demands of the embedding culture than are the "natural" sciences—if for no other reason than that the phenomena they address are themselves embedded in the culture. One specific reason is that there are two possible sources of social/historical influences on the structure and content of psychological research. One is the influence of the social context on the possible topics and theories that a researcher/theorist may think about and act on, the other is a function of the fact that practically all our research on human psychology is conducted with products of our culture and therefore the behavior of our research subjects is to some (unknown) extent a social product.

If we accept that the sciences, and the social sciences in particular, are to some extent culture driven, is it possible that we may still discern some progress, some advancement in our understanding of human thought and action?[3] Depending on one's particular orientation, one might see laudable advancements in the history of psychology, as in the institutionalization of the experimental method by Wundt and his peers or in the displacement of the associationism of the nineteenth century by labeled associations and organizational structures. Changes, however, are clothed in the contemporary cultural values. Wundt reflects the ambivalences of his and Germany's history, Gestalt psychology mirrors the changes in the visual and intellectual culture of the turn of the century, behaviorism embodies the parochial and puritanic concerns of early twentieth-century America, and current cognitive psychology is a product of the information and communication revolution after World War II. I move now to an illustration of these claims.

Choicepoints in the History of Psychology

I start with the case history of Wundt, only in part because of his traditional role as the "founder" of experimental psychology. My interest in Wundt is rooted in his intellectual background, the duality of his interests, and his influence on succeeding generations. Following that exposition, I discuss some developments in the early and mid-twentieth century within a wider social context.

Wilhelm Wundt: The German Ambivalences

I present three aspects of Wundt: his personal and political history, the social and political background in which he worked, and his contribution to two quite different psychologies. The story points to interesting co-occurrences in the three tales which by themselves cannot be taken to indicate any causal chains. However, these parallels are illuminating and at least indicative of the common context of personal, scientific, and social forces.

Wundt's contribution to the establishment of an independent psychological science came toward the end of an epochal period of fundamental change in European natural and social science. The mid-nineteenth century saw the contributions of Darwin and Marx, who changed knowledge about and attitudes toward human origins and the causes of our economic behavior.[4] Major changes occurred in most of the natural sciences, which led one commentator toward the end of the century to assert that physics was "completed." By moving toward independence from the conceptual umbrella of philosophy, psychology took a similar step forward. Below, I discuss Wundt's personal and academic history, followed by a general discussion of the situation of the sciences in the Germany of his time.

WUNDT'S LIFE. Wundt was born in 1832 in the small town of Neckarau; his father was a country pastor from an academic family. Young Wilhelm attended the Gymnasium in Heidelberg, and at the age of 16 he was an active and romantic supporter of the 1848 revolution. He enthusiastically read Georg Herwegh, the poet of the German Revolution of 1848, supported the brief republican uprisings during that period, and later regretted the loss into emigration of their leaders and the period of reaction that followed in the 1850s.[5] After attending the Gymnasium, Wundt undertook medical studies in Tübingen and Heidelberg, which he completed in 1855 when he was 23 years old. He stayed in Heidelberg until 1874, except for a year in Berlin and occasional periods in Karlsruhe to pursue his temporary political career.

With the importance of liberal politics in Germany after 1848 and in particular during their dominance in the 1860s, the liberals' commitment to a marriage of politics and education motivated scientists and humanists to combine scholarship and politics.[6] In Heidelberg, Wundt and many of his academic colleagues became active in the workers' education movement, and he lectured in Heidelberg and surrounding towns on Darwin and other topics of general interest. With the 1860s the workers' associations became transformed, in part under the influence of Ferdinand Lassalle and Marx, into more militant organizations that rejected their bourgeois leaders and lecturers. Wundt, who had enjoyed some of his interactions with prominent social democrats such as August Bebel, regretted this shift in the workers' movement and turned toward the center/liberal bourgeois organizations, becoming a member of the *Ständekammer* (representing the various petit bourgeois guilds and professions). He eventually became a candidate of the Progress Party and was elected a member of the second legislative chamber of Baden. There followed four years of great activity in the state capital of Karlsruhe. In the conflict between Prussia and Austria, most of the Baden population supported Austria. Following the defeat of the Austrians in 1866, the situation in Baden was very much up in the air, with Bismarck delaying the admission of Baden to the North German Confederation, but rejecting all advances from pro-French directions. Wundt found the situation "unbearable" and resigned from the legislature in 1868 to turn, once and for all, from politics and to devote all his time to academic pursuits. Two years later, in 1870, following the defeat of France by Prussia and her allies, Germany was united under Prussian leadership.

After attaining his medical degree in 1855, Wundt worked briefly in the Heidelberg medical clinic, but in 1856 he moved to the university in Berlin where he pursued studies toward his goal of doing research in general physiology. He benefited from the physiological research there under the direction of Johannes Müller and Emil Du Bois-Reymond. However, Wundt moved back to Heidelberg in order to achieve faculty status *(Habilitation)*, but soon afterward became very ill. He did finish a book on muscle movement, which was published in 1858 but was generally ignored. In 1858 the eminent physicist and physiologist Hermann Helmholtz moved to Heidelberg, and Wundt became his assistant—a relatively unproductive collaboration that Wundt gave up to write textbooks on general physiology.

During his early period in Heidelberg, Wundt saw sensory physiology as his true avocation and soon started work on his *Principles of Physiological Psychology,* the *Grundzüge,* which was published in 1874, the year he left the university. Wundt considered the succeeding editions of his work and particularly the sixth edition of three volumes (1908–1911) to represent a substantial part of his life's work.[7] He also developed the justification and basis for the experimental psychology that he was to father, insisting that whereas it was originally called "physiological" because of his background in physiological work, he considered "experimental" not to be restricted to the natural sciences but an important part of an exact social science *(Geisteswissenschaft)* which combined objective methods with self-observation.[8]

At the time, the social sciences, and psychology in particular, were preoccupied with the distinction, introduced by Dilthey in 1883, between the natural and social (mental) science—*Naturwissenschaften* and *Geisteswissenschaften*.[9] Dilthey actually makes two major arguments in distinguishing between sciences of the mind *(Geist)* and of nature. He notes that there are two possible perspectives: One can either regard everything from the perspective of the natural sciences, or one can regard natural phenomena from the perspective of consciousness. He believed, however, that our physical bodies are governed by the laws of nature. Thus, whereas "purpose" is a mental phenomenon, it is "physically" realized in the systems of natural phenomena and laws. On the other hand, Dilthey claimed a more primitive, basic status for the mental sciences since they are in closer contact with the "life-nexus" of our experience than are the natural sciences.[10] He also argues for the incomparability of material and mental processes based on the impossibility of deriving mental facts from those of the mechanical order of nature.[11] It was during this period of concern with the nature of the mental sciences that Wundt first reconsidered his interest in social psychological matters, which presented a task "higher" than experimental psychology. Wundt's 1863 book in that direction failed, but the thought was to be taken up later—postponed but not forgotten. The first volume of the *Völkerpsychologie* (on language) was not to appear until 1900.[12]

A fleeting year of Wundt's first full professorship in Zurich from 1874 was followed by the offer of a professorship in Leipzig 1875. He quickly adapted to this new environment and made the acquaintance of the two men who influenced his psychological work more than any others: E. H. Weber and G. T. Fechner, the two fathers of psychophysics.[13] In 1879 Wundt moved into a few rooms that constituted the Institute of Psychology and was given a paid assistant. The institute grew and moved repeatedly—in 1897 to more or less permanent quarters.[14] During the remaining years, until 1917, Wundt maintained his interest in philosophy and its history as he directed the new psychology. And starting in 1900 he published the 10 volumes of his other ("higher") psychology, the *Völkerpsychologie*.

By the time of his death in 1920, when Wundt's autobiography appeared, the liberal radical of his youth had become the nationalist reactionary of his old age. He assigned responsibility for Germany's defeat in the First World War entirely to the social democratic and other liberal opposition (the *Dolchstoss* position; i.e., Germany was defeated from within) and he assigned principal responsibility for the war to the deliberate planning by England and its leading statesmen, under influence of Benthamite utilitarianism.[15]

WUNDT'S GERMANY. In Germany the development of an empirical psychology had followed the emergence of a strong industrial basis in the middle of the nineteenth century, generally supported, subsidized, and encouraged by the state in the "education for industry" movement. By the end of the century large numbers of German youth were being channelled into the universities, and the end of the century also saw a more general technological advance. The late 1800s had produced the major steps of the second industrial revolution (sometimes called the new technology), including the telephone, wireless telegraphy, the airplane, the development of the German chemical industry (driven in part by Fritz Haber's invention of nitrogen-fixing), the diesel engine, and so forth. With the new technology and the newly educated masses came new attitudes in many fields, including the sciences, as well as a new psychology.

Johannes Müller, a true polymath, was the towering figure in German science during the mid-nineteenth century and was the guiding spirit in moving German academic theory and research away from *Naturphilosophie* and rationalism toward an experientially based, empirical natural science—from an a priori philosophy of nature to a naturalistic science of nature. Müller created the *Berliner Schule* and brought German science up to the empirical standards previously exemplified by England and France.[16] The movement rejected the speculative, vitalistic, and idealistic *Naturphilosophie* as exemplified by Hegel and Schelling. Though focused on the role of science in the universities, the movement was part of a general mid-nineteenth century tendency toward a materialist interpretation as opposed to a Kantian/Hegelian idealist interpretation of history and nature. In the universities that battle was carried out to a large extent by Müller's students Du Bois-Reymond, Virchow, and Haeckel.

Once the movement for a naturalistic, experiential science began to pervade much of German thought, it created another ambivalence. A backlash, particularly among its popularizers, was created against science and scientism. This counterflow tried to denigrate science and elevate a vitalistic philosophical, and often blindly destructive, approach in its stead.[17] In part this was a reaction against a hyperpositivistic point of view created by some of the scientists themselves— a fact-oriented approach that had rejected speculation and, as a consequence, also theoretical thought.

Wundt had benefited from his acquaintance and common interest with Müller, Helmholtz, Du Bois-Reymond, and the young physicists around them. He was introduced early in his career to a point of view that was to replace the traditional idealism and rationalism of German psychology with a materialistic/empiricist point of view. Another movement in the same direction that affected the emergence of a quantitative experimental psychology was the introduction of statistical thinking and procedures in the latter half of the nineteenth century.[18] These influences represented another tension in Wundt's life and career—though not one he easily acknowledged. The young Turks in science were materialists, whereas Wundt was a life-long idealist, suspicious of materialist doctrines. He committed psychology both to a *Geisteswissenschaft* and to an experimental science. By this step he avoided the materialism of the *Naturwissenschaften* while at the same time embracing an experimental empiricism.[19] The underlying argument incorporated something like the following reasoning: Laws, like the laws of natural science, were not possible for the complexities of mental life. But if a reductive materialism were to be defended, then thoughts and emotions could be considered to be just collections of neurons and nervous impulses, and the laws of mental life would be natural science laws. Given that there are no such laws of mental life, materialism is indefensible; psychology is a *Geisteswissenschaft*, not a *Naturwissenschaft*. And furthermore, it is relatively easy to defend experimentalism from an antimaterialist point of view, since experiments can reveal only sufficient, but not necessary, conditions.[20] A hundred years later, with more sophisticated though not necessarily "better" views of the relations between materialism and science, the elusive hunt for a reductionist, materialist position continues as always in vain.[21] At the same time, given the elusiveness of a strict materialist position, the distinction between natu-ral/materialist and mental/antimaterialist sciences has lost both its validity and its force.

The success in turning psychology from rationalist philosophy to empirical science, which was originally opposed by traditional philosophers, turned philosophy full circle as, in the early twentieth century, traditional philosophers objected to the inclusion of experimental psychologists in their departments. The creation of independent psychology departments was a natural consequence.

With the failure of the ideals of the 1848 revolution, particularly after 1866, the young scientists turned away from politics. Instead, they hoped and worked for the reform of the authoritarian structure of the German universities, in part to stress the place and importance of natural science. Wundt was younger than these men, and whereas he apparently absorbed many of their attitudes, he showed little inclination to challenge university authorities and regulations. The failure of the hopes of 1848, followed by the triumphs of Prussian hegemony in 1866 and 1871, foreshadowed the general shift in Germany from liberalism to nationalism. Parallel developments tended to suppress the democratic/liberal tendencies in Baden and other states and to terminate their independence. In 1871, following the defeat of France, Bismarck achieved the unification of the German states and the establishment of the Second Reich. By that time Wundt had essentially changed sides and supported the united Germany under Prussia. He reflected the change in the attitudes of the professoriat that became complete by the end of the century. The scholar was to remain on the sidelines of politics and neutral toward its opinions and interests.[22] A German banker returning in 1886 after a long absence noted the rampant militarism and observed that university professors, the former leaders of liberalism, "now kowtowed to the authorities in the most servile manner."[23]

Over the span of Wundt's life, Germany (and the province of Baden in particular) had changed from a democratic/liberal to an authoritarian/militaristic state. Wundt's personal politics had followed a similar path. Democratic Germany was emergent in 1848, the 1920s, and again in the 1960s; the authoritarian strand dominated German history from the 1870s until after the First World War and again in the 1930s and 1940s. In a changing Germany with its varied and ambivalent attitudes, Wundt grew up and matured. This pattern was reflected in Wundt's own transformation from a liberal democrat to a Prussianized older man.

WUNDT'S PSYCHOLOGIES. The duality of German politics and the parallel development of Wundt's politics are reflected in his psychology, which has produced a split between social and experimental psychology and has been with us in some guise or another ever since. I do not wish to imply a one-to-one mapping of German politics, Wundt's life, and his psychology. Rather, I stress the parallel ambivalences present in all three of these domains. They were reflected at various times in Wundt's psychology, so that, for example, his interest in social phenomena can be seen through much of his life, and his concern with rationality and order, though strongest at the end of his life, was present even in his early "liberal" days. Wundt made a strict distinction between experimental psychology on the one hand and ethnopsychological and social psychological topics that were not subject to experimental investigation on the other hand. The latter were to be approached in a rational and observational manner, whereas experimental psychology was part of the experimental sciences. I argue that this distinction reflects in part the duality in German history, between a democratic, humanitarian tendency on one hand, and an authoritarian one on the other.[24] The inquiry into development of the human mind—the social dimension—allowed Wundt to present his humanist, social concerns and to be concerned with the contingencies rather than with the rules of behavior. Experimental psychology was to a large extent sensory psychology. It was strictly scientific, following rigid rules of experimentation, and did not allow any "softer" concerns. In the same vein, and in part due to the influence of Fechner, Wundt adopted statistical error theory in experimental psychology, though he rejected statistical laws for the historical phenomena treated in the *Völkerpsychologie*.[25]

Wundt's topics in experimental psychology were sensory processes, perception, consciousness, attention, will, affect, time and space perception—the classical topics of psychology. He viewed the more complex phenomena as built up from pure sensations and feelings *(Empfindungen und Gefühle)*. Whereas his scientific/academic books restricted psychology to its possible experimental foundations, they often introduced the topics of his *Völkerpsychologie, Sprache, Mythus, Sitte* (language, mythology, culture), without dealing with them at length.[26]

Starting early in his career (around 1860) Wundt had been concerned with the "higher" psychological functions, the social aspect of human thought and behavior, and this preoccupied him particularly in the last 20 years of his life. The multivolume *Völkerpsychologie* illustrated Wundt's commitment to the notion that the only source of insight into the development of human thought can be found in its social/historical development. His method was an extensive use of existing ethnographic and anthropological writing, as well as the most widely used extrapolations on language, the family, and other human institutions. It was written from the point of view of a comparative evolutionary perspective; Wundt was particularly concerned with the development of communication systems starting with a gesture language, then developing into "vocal" gestures and then into language. He related the three broad topics *Sprache, Mythus, and Sitte* to the individual-psychological aspects of representation, emotion, will, and habit. All these topics were seen by Wundt to be tied to sociality and the historical conditions which give rise to it. The purpose was to deal with "those psychological processes that form the basis of the general development of human societies as well as of the creation of common mental products of general validity."[27] In his effort to understand mental structure and development across the development of culture and language, Wundt defended this function as against anthropological concerns. He noted that whereas psychology must be based on the results of ethnology/anthropology, mental development may still be the same for different cultures or that similar cultures may, psychologically speaking, represent different stages of "mental culture." In his quasi-summary of the *Völkerpsychologie*[28] he discusses in detail the early development of human societies, the acquisition of gesture and verbal language (where he first introduced phrase-structure analyses), the role and development of marriage, myths, and religion. Although his approach was marked by some of the ethnocentric prejudices of the nineteenth century, the analysis of social and cultural behavior and thought was novel and clearly distinguished from the other, experimental, psychology.

Wundt's distinction between experimental psychology and ethnopsychology was in part in the tradition of Dilthey's distinction between *Geisteswissenschaft* and *Naturwissenschaft* (mental versus natural sciences). Wundt produced an experimental psychology that was not social and a social psychology that was not experimental. His experimental psychology defined Western psychology, but his social psychology had little influence in psychology (in Germany or elsewhere), though some sociologists like Durkheim were significantly influenced by him. Important in this lack of effects is the fact that only a short one-volume extract of the *Völkerpsychologie* has appeared in English.[29] However, the distinction he made between the two psychologies is still with us. Experimental and social psychologists in the United States, though often trained side by side, often take only superficial note of each other's research (some notable exceptions notwithstanding). Even a philosopher of cognitive science recently excluded consciousness from being efficacious in cognition but still possibly relevant to resolving conflicts and for planning or for pedagogical or social purposes.[30] It is difficult in this context not to absorb an attitude that makes "cognition," however defined, somehow quite different and apart from social concerns and behavior.

To summarize, I stress Wundt's ideas and directions for several reasons. Maybe most important among these is the fact that practically all influential psychologists at the turn of the century were students of Wundt's or were students of his students. Experimental psychology was defined by the experiences of the Leipzig laboratory, and American laboratories were generally opened by his students with imported German instruments. The other reason for stressing Wundt is the duality of his approach. The split that characterized Germany in the nineteenth century between democratic, and even radical, idealism on the one hand and Prussian rational militarism on the other was echoed in Wundt's life and, most important, in his psychology. The contrast was between an experimental, tightly reasoned, mathematical psychology and a historical, observational attitude toward those elaborate products of the mind that generate language and culture but which are beyond measurement and beyond experiments. Wundt preached the inaccessibility of those phenomena to experimental and by implication "scientific" study. He asserted their dependence on historical and social influences and determinants. Experimental psychologists sometimes have used the distinction for promoting their exalted status as scientists, whereas social psychologists remained ambivalent about whether they were "scientists" or could follow Wundt's road into social analysis.

Interlude: Williams James and Functionalism

I refer to the period starting about 1880 and ending in the early 1900s as an interlude because it was concerned to a large extent with cleaning up the loose ends and leftovers of the nineteenth century. William James was the great summarizer and only to a lesser extent a theorist, and the great period of the British empiricists/associationists came to an end. At the same time, Edward Titchener was playing out a rather idiosyncratic version of Wundt's psychology, which left no discernible important traces. During this period major changes were taking place in the intellectual life in Europe which were reflected in psychology in the next century, which I discuss in the next section. However, I fail to find any obvious cultural trends or any specific links between the predominant culture and the contemporary themes of psychology. I believe that both U.S. and European history and theoretical psychology were in a transitional phase of consolidation in which no major themes emerged.

The other great influence in the late nineteenth century, next to Wundt, was James, who only for part of his productive and creative life was a psychologist, but he set the tone for American psychology for decades to come with his *Principles of Psychology,* published in 1890. The *Principles* took some 12 years to write, as they incorporated most of the seminal papers and reviews James had written in the years preceding 1890. James was not a great systematizer; the book is a rambling collection of thoughts, facts and speculations, but it was encyclopedic, well written, and imbued with the American spirit of keeping contact with the everyday practical world. James' novel contributions were his work on emotion and on habit (the "great flywheel of society"), the developments of the notion of the "stream of consciousness," and an analysis of memory that has survived to the present. He made psychology exciting and accessible, not least by his humanity, knowledge, and writing skills. But James was entirely dedicated to an individual psychology; there are no considerations of the human mind molded by society. In a sense he mirrors the individualism of the American frontier. For James, one moved from physiology to individual psychology, which was a natural science. For Wundt, the road branched after physiology to an experimental psychology that was a *Geisteswissenschaft* and to a social science of the human mind situated in history and society. Both of them were dedicated to bringing about a sharp break with the preceding philosophers of psychology, their rationalism and their abstract theorizing. Both were committed to empiricism, though Wundt's psychology was voluntarist, whereas James was less concerned with the will and more concerned with the wheel of habit.[31]

In the United States the commitment to empiricism was one of the new directions of psychology, and an acknowledgment of the influence of Darwinian ideas was the other. While Darwinian thought in the first few decades was often characterized by wild speculation that would put even some contemporary sociobiologists to shame, it did build a slow and solid concern with evolutionary matters in the psychological community. James was appalled by Herbert Spencer and social Darwinism but represented the beginning of the earliest American biologically (not physiologically) oriented movement—American functionalism. These functionalist psychologists, starting with John Dewey and including Carr, Angell, and others, were concerned with the evolutionary functions of behavior[32] and the establishment of empirical relationships. The latter were primarily concerned with looking for empirical laws, literally the observation of and experimentation with behavior as "a function of x." Their last important contribution was in 1942 with McGeoch's book on human learning with chapter headings that actually were "x as a function of y."

In the United States, psychology in the late nineteenth century was dominated by Wundt's students, and in the years following the pilgrimage to Leipzig, there emerged a dominant figure in America—the experimental introspectionist E. B. Titchener. Titchener is primarily responsible for identifying Wundt with a dreary psychology concerned with "introspecting" the contents of mind. Titchener's introspectionists, whose training defined what could be observed, proclaimed the "observed" mental contents to be the building blocks of mind.[33] Sheer phenomenal experience was the stuff of psychology, and in the process Titchener identified some hundreds of individual experiential elements that structured mental contents. Titchener prepared the ground for the birth of behaviorism in America in the second decade of the new century. The infertility of much of Titchenerian psychology marks the end of the introspective and atomistic tradition of experimental psychology, while the Wundtian thought on social and cultural phenomena never effectively reached the North American continent—nor did it have any significant following in Europe.

With the advent of a respectable experimental psychology and with the opening of the major psychological laboratories in the United States, the nineteenth century came to an end and the new century introduced modernism in its first decade—a modernism that would manifest itself quite differently in the United States and in Germany—with the advent of behaviorism and Gestalt psychology.[34]

Technology, Society, and Psychology in the Twentieth Century

Wundt epitomized the new directions for psychology in the nineteenth century. With the new century a number of different, and sometimes more important, changes would take place in the definition and direction of psychology. Psychology broke with much of its past, just as both European and American society and culture in general exhibited new directions in the twentieth century.

The new century brought with it the beginnings of new developments in the arts and sciences. Whatever the manifest reasons for the outbreak of World War I, the breakdown of old alliances and the unavoidable conflict among the major Western powers apparently created the conditions for changes in areas other than big-power relations and commercial and industrial realignments.[35] With the twentieth century came important breaks in art and architecture with the arrival of the *Jugendstil* and art nouveau and their rejection of the imitative styles of the nineteenth century. In music new modes of expression were found in the work of Schönberg and others, again a sign of the rejection of the old. In France, Matisse and the Fauves, followed by the Cubists, also rejected a sheer or mere presentation of nature. And Joyce opened new doors of expression in literature.

The last decade of the century saw a cluster of genius emerging in the intellectual life of Europe that would act on intellectual life well into the 1930s. H. Stuart Hughes notes such a "cluster" and drew attention to the fact that an amazing group of thinkers was born in the period between the mid 1850s and 1870s, including Freud, Durkheim, Bergson, Weber, Croce, Pirandello, Gide, Proust, Jung, Mann, and Hesse.[36] All of them participated in the change from a view of humanity "as self-consciously rational" beings to a more limited, irrational (unconscious?) view of human motives and human freedom.[37]

In the United States, the new century saw basic changes of values in American society. The new direction was practical, bureaucratic, and concerned with the rationalization of industry and the establishment of social order. It was also somewhat nationalistic, pragmatic, and inner-directed. American science was dominated by its pragmatism and its technological emphases, exemplified in the popular view of Thomas Edison (1847–1931) as its premier scientist.

In Europe the new century was to see important new directions in psychology. In the long run the atomistic/experiential psychology of Wundt was followed by the holistic/experiential psychologies of Selz and the Gestalt psychologists, whereas in America Titchener was followed by the atomistic/objectivist tradition of behaviorism. As I note below, this rather surprising bifurcation was, at least in part, a function of new ways of looking at the world in Europe and new emphases on inward-looking strands in the United States.[38]

More important, however, was the development of a psychological theoretical language *sui generis*. It was another expression of the rejection of a sheer mirroring of nature and of common sense. Psychology participated in the rejection of the "self-conscious rational" nineteenth-century image. When Freud rejected the notion that human motives are self-evident, he needed a new, psychological language to express this new direction. The general cultural movements toward new ways of expression in art, music, and literature had their parallel in the movements toward a psychological language. Theoretical speculation, both in Wundt and James, usually consisted of rather vague psychological concepts but a commitment to the notion that true psychological explanation was to be found in the physiological substratum (of the day).[39]

Producing a theoretical language implies the postulation of theoretical entities and processes—and if they were not explicitly neurophysiological ones, then one needed what came to be called "unconscious mental mechanisms." Whereas in the last quarter of the nineteenth century it was still generally accepted that most or all the important mental events were conscious, by the last quarter of the twentieth century we now accept that most of the important determining events are unconscious. This shift was part of a general trend toward abandoning the view of human rationality (viz. Freud) or toward a greater emphasis on theory and its language, or, more likely, both.

Two unrelated developments led to the emergence of a psychological language. One was Freud's invocation of the unconscious, though not its "discovery."[40] By assigning to the unconscious psychological functions akin to conscious wishing, willing, and avoiding, Freud invoked a theoretical language that operated as the underlying representation of overt thought and behavior. At the same time, a series of experiments coming out of Würzburg, in retrospect somewhat unconvincing by themselves, generated a similar need for a layer of theoretical concepts that were neither conscious contents nor physiological entities.[41] The "discovery" was that thinking was sometimes unconscious; when solving problems, producing associations, and making judgments, the solutions, judgments, and the meanings of concepts and words were often given directly—i.e., without intervening conscious contents. The term in English, "imageless thought," which has been given to these investigations is relevant for some instances that lack accompanying imagery, but generally the more appropriate reference is to "unconscious thought." And it was rapidly followed by the introduction of new categories of consciousness such as conscious dispositions (*Bewusstseinslagen*), Ach's *Bewusstheit* (essentially the modern metacognitions), and the notion of the unconsciously registered *Aufgabe* or task. Within a few years the consensus was that there was a paucity of conscious evidence for complex thought processes. The same conclusion was reached by Binet in 1903 when he noted that one gets a nickel's worth of conscious images for a million dollars worth of thought.[42]

The intellectual turmoil of the first quarter of the twentieth century was reflected in both pre- and post-World War I changes in culture and society. In Germany the changes in the art world from *Jugendstil* to expressionism also were mirrored in similar changes in literature and popular culture. In psychology it was reflected in psychology dominance of Gestalt psychology and its insistence on structure and the power of the environment. The discontinuity between Wundt and the new psychology reflected the change to twentieth-century modernism. Gestalt ideas ranged far beyond their concerns with perception as they were deployed in Köhler's work on problem solving with primates, Koffka's concern with development, Wertheimer's and Duncker's contribution to human problem solving, Katona's contribution to the organization of human memory (to be rediscovered in the United States in the 1960s), and, of course, Lewin's contribution to social psychology.

Somewhat buried in the success of Gestalt psychology was the work of Otto Selz, a German psychologist who developed a psychology of thought even more consistent with contemporary work than Gestalt psychology.[43] Selz was concerned with processes rather than contents of thought and in particular with productive and reproductive thinking within a single system.[44] He was very much the forerunner of contemporary modelers concerned with operations on knowledge structures.

America and Behaviorism

In contrast to German industrialism and Prussian authoritarianism, tinged with dreams of empire and romanticism, American society in the early twentieth century was inward looking, marked by the unique American experience of the expanding frontier and the aftermaths of the Civil War and imperialist expansion. Internationally, the forces of growth and expansion had become loud and insistent at the end of the century. After the Civil War had reduced the remaining feudal concerns of American society, American industry—and isolationism— were preoccupied with building an expanding country, and the end of the century saw the beginning of American expansionism and imperial ambitions. The building of a world-class navy, the annexation of Hawaii, and the application of the Monroe Doctrine all contributed to a sense of destiny, which culminated in the annexation of the Philippines. The latter is important because older American values were articulated in opposition to the Pacific expansion. The war with Spain started in 1898 over Cuba. Soon thereafter the Anti-Imperialist League, which united some of the finest American public figures, joined the fight in opposition to the war. However, the opposition movement did not succeed, and Hawaii and the Philippines became American. During the battle in and out of Congress, appeals to the inheritors of the ideals of the American Revolution failed. James commented that "the way the country puked up its ancient principles at the first touch of temptation was sickening," and that America was "engaged in crushing out the sacredest thing in this great human world— the attempt of a people long enslaved" to work out its own destiny.[45] The partial success of the imperial adventure and the fissure it revealed in American attitudes on expansion and colonialism led to a more inward-looking atmosphere, in part by restructuring the internal state and in part by producing an intellectual isolationism. Politically, the new century brought a consolidation of American imperialism, and a look south rather than across the seas. In science and philosophy the period was marked by a pragmatic, untheoretical preoccupation with making things work—a trend that found its expression in psychology in functionalism and behaviorism.

The inward-looking period of the turn of the century was reflected in part in the parochialism and the environmentalism of the behaviorist movement. Behaviorism had also responded to the drive of the new technology in the early 1900s. The new movement in America—the revamping of its values—was in part a parallel to the deliberate German encouragement of industry in the early nineteenth century. In the United States a new middle class of urban professionals developed the values of "continuity and regularity, functionality and rationality, administration, and management" in response to the sense of disorganization following the end of reconstruction; what was needed was "government of continuous involvement"[46] that would make possible the further expansion of American capitalism and a sense of order, some control over the economic and social forces. Behaviorism was partly a response to this new set of values, and it was also a very American response to the inherited attitudes of the Puritan ethic. Watson set the stage in 1913 when he advocated a "purely objective" method for the "control of behavior."[47] Behaviorism rejected Titchener and the German tradition as being irrelevant to the daily concern of people, and it would have nothing to do with the elaborate theoretical fictions arising in Germany and France. It was a psychology stripped to the bare essentials—pragmatic, at times anti-intellectual, self-consciously Puritan American. Like the reformist response to Catholicism, the fancy, complex elaborations of a basic faith were rejected: Behaviorism was a call for basics. Watson's manifestos rejected analyses of consciousness and theoretical invocations and stuck with observables, both in the environment and in behavior. Thought was subliminal speech and, particularly with the adoption of the Pavlovian approach, complex behavior was to be understood from the combination of simple behavioral rules—true of all mammals within the context of a simplistic evolutionary perspective. Watson directly tied behaviorism to the wider American concerns by stressing behaviorism's native character, that it was "purely an American production."[48] Esper notes how extraordinary that statement was, coming from a pupil of the (German-born) protobehaviorist Jacques Loeb, and a student of Jennings and (Britisher) Sherrington who picked up much of his experimental method from (the Russians) Pavlov and Bekhterev. So much for the "myth of the immaculate conception of American behaviorism."[49]

The initial success of behaviorism spawned a variety of different strands and directions. Central to nearly all of them was a preoccupation with simple mammalian rules that enthroned *mus norvegicus* as the major focus of research. In contrast to the radical behaviorists, some theoretical entities entered the fold, including cognitive ones in Tolman's influential cognitivist labors, deviant from the dominant associationist thinking. But even when theory was invoked, it never strayed far from a language tied to stimuli and responses, as in Hull's ambitious, but in the end failed, system.[50]

It may still be too early to determine which of the various strands that influenced the development of behaviorism was most important. Suffice it to say that the movement itself was consistent with a number of old and new American cultural and social values.[51] It also had its kindred movements in such developments as the drive for scientific management and the time-and-motion studies of F. W. Taylor designed to make the American worker more productive at less cost.[52]

World War II: A New Mode of Life Emerges

By the 1920s and 1930s the new developments were fairly stable, whether it was Gestalt psychology in Germany, behaviorism in the United States, or Pavlovian psychology in the Soviet Union. But soon radical changes took place in the overture to World War II. The Gestalt psychologists were driven from Germany, and most settled in the United States, where they were relegated to minor positions in universities and colleges at the edges of the American establishment.[53] Behaviorism enunciated the aspirations of American psychology for a single encompassing system that would eventually understand all of human and animal behavior. However, we now arrive at the academic pause impelled by the war from 1939 to 1945. The war quieted any serious theoretical activity as the most active people, like most others, were absorbed in applied work in support of military needs and goals.

The post-war changes in culture and technology came relatively slowly but were no less drastic than previous ones. In technology at large, Turing prepared the ground for the computer age and von Neumann imagined the digital computer, the dormant development of television became a reality, and radar and its related cousins, born in the war, contributed their novelties. The changes accelerated with the invention of the transistor in 1947 and the later development of integrated circuits.

With the end of war, the information revolution, tentatively started in the 1930s, became a reality, and both industry and the scientific/intellectual establishment exhibited the symptoms of a new way of social organization centered on knowledge acquisition and transmission. Shannon and Weaver popularized information theoretical analyses,[54] von Neumann, Morgenstern, and others produced notions of computation and game theory, and other ideas from engineering and physics staked claims to an understanding of human behavior—including, of course, Norbert Wiener's cybernetics. The information revolution had begun, but not quite in psychology.

Since all major cultural changes are overdetermined, I shall try to do justice to various strands that signaled the change in psychological research and theory. At this point one cannot give pride of place to any one of these developments; we are too close in time to those changes to decide which ones were central and which were peripheral.

One such strand can be found in the midst of the military activities of the war. The war period had contributed in part to the changes in psychological theory and research. The war effort brought together a number of people in a number of efforts. Of special interest to later developments was a group at Harvard, which included J. C. R. Licklider, S. S. Stevens, Ira Hirsh, Walter Rosenblith, George A. Miller, W. R. Garner, and Clifford Morgan. Their primary concern was in psychoacoustics and noise research, but it also extended into signal detection and related topics. Similar accumulations of talent occurred in other parts of the U.S. war establishment as well as in Britain (e.g., in the influence of the work on vigilance of D. E. Broadbent).

The opposition to the behaviorist hegemony in the United States had been carried mainly by the Gestalt psychologists and their allies. But in 1949 there appeared a major alternative to the mainstream of behaviorism. D. O. Hebb's *The Organization of Behavior* was in part ignored, to the eventual embarrassment of the conventional wisdom, but found enough support to become the core of a small counterrevolutionary movement. Hebb introduced notions of organization (i.e., organized rather than atomistic stimulus–response chains) but also cut the knot between physiology and psychology.[55] Being a physiologist by "birth," Hebb coined the term "CNS" (the conceptual nervous system), noting that the invocation of quasi-physiological concepts, such as cell assemblies, was not necessarily a commitment to physiological reduction but rather a halfway house between the disciplines, where such concepts could be used as theoretical terms by psychologists. In the United States it was another half dozen years before the rejection of behaviorist dicta hit full stride. A burst of activity rarely seen before turned the field around between 1955 and 1960 and established a firm basis for the "new" cognitive psychology.[56] In the United States, the rejection of behaviorism during that period had the flavor of a counterrevolutionary movement, with all the perquisite fervor displayed on both sides.

The development toward a basically theoretical (cognitive) psychology started earlier in Great Britain with somewhat less immediate general influence. This lack of influence may in part be due to the lack of a highly motivated opponent in the behaviorist tradition, which was essentially absent in Great Britain. The British work is seen in the experimental and theoretical work of Bartlett, the brilliant anticipation of cognitive modeling by the prematurely deceased Kenneth Craik, and by the subsequent development of these trends by Broadbent.[57]

The contribution of French psychology has been largely overlooked, which is unfortunate because the work of Edouard Claparède, Binet and others is very much related to modern cognitive approaches. For example, in the same two decades that Köhler published his work on insightful solutions, Claparède was studying problem solving, culminating in the notion of the hypothesis as central to problem solving, and incidentally producing psychology's first protocol analyses.[58] The important contribution of the francophone Jean Piaget was also delayed at the international level until the 1960s, when his innovative use of the concept of the schema became important in cognitive psychology. However, the mutual isolation of French and American psychology prevented the significant influence that the French psychologists deserved. They too, of course, had no behaviorist antagonists.[59]

Another discernible strand in the development of the new psychology was the resurrection of old conceptual directions that had been dormant. Apart from organization, there was the notion of the schema and of "cognition" itself. The schema concept,[60] originally introduced by Immanuel Kant, has remained essentially unchanged in its use by Piaget, its implication in Bartlett's work, and in its current incarnation.[61] The schema has not only become scientifically respectable, but it has also provided a link to connectionism. The term "cognitive psychology" had currency before its current incarnation, but it was seen as a fuzzy, vague approach, often categorized (probably incorrectly) with personalistic, phenomenological approaches.[62]

All of these strands and tendencies produced the setting for the major changes in experimental and academic psychology. New and old ideas now found fertile ground for their development. The importance of the right time and the right place in these cultural changes can be illustrated by a couple of case histories. Important and influential papers written several years earlier did not make an impact until "the time was ripe." Lashley's paper on serial order, with its rejection of contemporary associationism, published in 1951, did not have any significant following until about 1960. Bruce has provided an insightful account of the fate of Lashley's paper and of the 1948 Hixon symposium, which occurred about 10 years too early.[63] He notes that the Lashley serial order paper did not energize the new trends, but rather supported and shaped them, and the same can be said of many other significant contributions of the time—they were all supporters; the energy came from larger forces in contemporary history. No single event or group of events is causally responsible for that period of activity.

Another example stems from the publication in 1940 of two books on memory, Katona's book on organization and the book of Hull and colleagues on an attempt to bring S-R psychology to rote learning. Katona's book was ignored as being irrelevant and beyond the reach of contemporary psychological science. The Hull volume was in the mainstream of psychology but conceptually and methodologically useless. As a result, neither had any effect at the time, but Katona was rediscovered in the 1960s when U.S. memory psychologists "discovered" the importance of organization. The lessons seem to be that one can be ignored by being "out of time" or by being "wrong"; being "right" is useless until the culture is ready.

The major events of the miraculous five years between 1955 and 1960 occurred in memory, language, and problem solving. Much of the activity was a recapitulation of the missing theoretical years. Psychologists rediscovered the British psychologist Bartlett and the importance of schemas and hypotheses in human thought and began to read Piaget as something other than a peripheral Swiss interested in the irrelevant behavior of his children. Also in Britain, Broadbent made attention a respectable word rather than—as in the United States—a hidden metaphor for the forbidden consciousness and told us about the importance of communication. The problem of mental organization, forgotten for a few decades—was reawakened and became a major theme of the first decade of cognitive psychology. Piaget's work, translated late in its development[64] and ignored by the behaviorists, ignited the explosion of efforts in cognitive development which, ironically, replaced the behaviorist dominance of learning problems and became our major source of insights into human learning. Considerations of and preoccupations with structure on one hand and information processing in the wider sense on the other hand preoccupied other human sciences during the same period. This can be seen in Chomsky's transformational grammar, in early attempts at artificial intelligence, and in the exploration of kinship structures in cultural anthropology. Whatever causes may eventually be definitively identified as the energy behind this wide-ranging reorganization, it certainly cannot be said that any one field or any one investigator started the changes or even predated them.[65]

The choice of the term "cognitive" was in part fortuitous; for most of that period—and in keeping with the culture that fostered it—the new theories were information-processing theories, but "cognitive" had been around for a long time to describe a psychology that imagined an active human organism operating on the environment. The influx of the mathematical psychologists and of the computer made the new psychology much more hard headed, and by the mid-1960s when Miller and Jerome Bruner started the Center for Cognitive Studies at Harvard and when Neisser published his (then definitive) *Cognitive Psychology*, the name had been chosen and stuck.[66]

The Recent Present

We are now nearly half a century away from that particular change in direction. Is it time for a new one? In which direction do we go?

The claims of reductionism discussed in chapters 4, 5, and 11 have been raised again recent years. The increasing interest in brain–behavior relationships and the rise of neuropsychology both have contributed to a revival of the view that complex (mental) phenomena can be understood (reduced) in their entirety to more basic (physiological) ones. I have already noted the importance of cognitive neuroscience in building bridges instead of reducing phenomena.

Complexly related to the drive for the physical substratum is the development of *cognitive science*. It arose in part because of the increasing curiosity and the desire for better bridges among the various cognitive sciences, in part because shrinking funding opportunities suggested more appeal when two or more intellectual directions are combined, in part because the artificial intelligence community had difficulty in finding an intellectual/academic home, and in part because of a genuine dream for a science of knowledge, human and otherwise. Some 20 years after its various births, there still is no such thing as a core cognitive science. Depending where one looks, what departments one queries, and who one's friends are, the core of cognitive science will be asserted to be neurophysiology, psychology, artificial intelligence, linguistics, or some more vague concept like human–machine interaction or symbolic or connectionist modeling. The result may not have been cognitive science, but it has been exciting and scientifically fruitful. It has created a community of interests and increased interdisciplinary communication. But as of now there are still viable independent cognitive sciences such as neurophysiology, linguistics, and psychology that flourish with or without the cognitive science label or affiliation. It is difficult to say at this point where it will lead.

A significant change I can discern at present is that some psychological habitats, like many other social sciences, are following the postmodernist trend in art and politics. A side effect of postmodernism has been the view that we are arriving at the end point of various scientific endeavors. In recent years, a variety of philosophers and cognitive scientists have acted as if the millennium had arrived and the final model that intervenes between the brain and behavior has been found in the computer analogy.[67] A similar attitude that "history is done with" can be found in some postmodern millenarian assertions that the era of general theories of psychology (of even limited domains) is over and done with.[68] Both of these mutually exclusive positions act as if we have come to the end point, as if the usual scientific progression from "truth" to error to new "truths" had finally come to an end, and that now all is known.[69] It is not yet the case that psychologists are willing to abandon general principles, but there is a trend apparent that sees phenomena as self-contained, with explanations, representations, and forms collected from a variety of sources, with the text becoming the central concern. In psychology, Jerry Fodor's multimodularity view of the human mind with each module doing "its own thing" harks back to an old faculty psychology and appears to be an obvious postmodern symptom.[70] Similarly, some aspects of the "deconstructive" tendencies of connectionism and its rejection of general principles of thought or behavior exhibit symptoms of postmodernism, as does the general atomization of psychology into different and unconnected subfields. Relations among experimental cognitive psychology and neuropsychology on the one hand, and social psychology or clinical psychology on the other, exist primarily in the exchange of terminology, not in any substantive unitary structure. The psychophysics of vision and hearing, once the heartland of psychological science, have become highly successful and quasi-independent fields. All this might well be the result of success as new fields separated out, as fields of philosophy did in the preceding centuries. It might also be the case that in psychology and related fields the flight to postmodernist atomization is a reincarnation of the battle between *Naturwissenschaft* and *Geisteswissenschaft* of more than a century ago. If mental and social regularities are not, or cannot be, represented in laws that are simple or useful in the sense of the laws of the natural sciences, then some other method might be more useful. Postmodern story telling and sociologizing are the consequences of abandoning the search for regularities and general principles. Conversely, one might return to the *Völkerpsychologie*—the social psychology of Wundt—and consider a nonexperimental but rigorous science of mind and society.

Whether or not one wants to go in the direction of the *Völkerpsychologie,* it is still likely, given the history of science and the history of ideas and society, that another mode of thinking will follow postmodernism. In other words, if postmodernism is a transitional state, then the next set of changes should indeed be interesting in their novel ways of dissecting reality and synthesizing new ways of seeing mind and behavior.

Physics Envy and Doing Psychology without Experiments

We have seen how Wundt bifurcated psychology into its experimental and non-experimental sections. This step had two consequences: the establishment of a dominance order in psychology, with the experimentalists becoming cocks of the roost, and an attempt, which reached its peak by the middle of the twentieth century, by the nonexperimentalists to establish their experimental credentials.[71] Little thought was given to the possibility that Wundt's psychological dualism may have been at least partially right. The main reason was the intent for psychology as a whole to become a "science." And sciences were, by definition, experimental. This tendency was a general reflection of trying to emulate the most successful of the various sciences. Wundt had not given any indication how the nonexperimental psychological topics could be scientific without being experimental. The result was an experimental social psychology, the kind of social psychology (though certainly not all of it) so decried by its critics.[72] The sometimes desperate attempt to experimentalize social topics resulted in laboratory experiments that were sometimes rather poor caricatures of the social phenomena to be investigated. What was missed was the opportunity to adopt "scientific" models that were not experimental but certainly had perfectly sound scientific credentials. I am thinking specifically about such fields of endeavor as paleontology and astronomy. Astronomy in particular has some rather interesting parallels with nonexperimental psychological phenomena.

Astronomers deal with objects, such as planets, stars, and galaxies, that are each unique and also follow general laws. These objects exist in aggregations that are characterized by the fact that the interactions among them determine in part the features and "behavior" of the individual objects. Astronomers tend to find new objects that display characteristics not seen before, and they adjust their theories to take account of these new findings. In order to bring some structure into their endeavor, astronomers also survey the types of objects (such as stars) that they encounter, and such surveys establish typologies—categories of objects that have similar defining characteristics. There are no experiments in the sense of manipulating variables and observing their effects. The similarities of this kind of endeavor with social and personality psychology should be obvious by now. I believe we missed the boat on that one, but I do not believe that a belated acceptance of such a model would soon repair the situation. Astronomy is as effective as it is because its theories (and observations) have been accumulated, adjusted, repaired, corrected over many hundreds of years. Astronomy has also been, over the years, interdisciplinary, as it appropriated and used the insights of mathematics, physics, chemistry, geology and others. Such an accretion of knowledge cannot be replicated overnight or over years. All that might be possible is to take existing observation and existing theory (usually experimentally derived) and start, however tentatively, a kind of "psycho-astronomical" endeavor.

The other model I mentioned has much in common with a "humanistic" field. Paleontology, and more specifically paleobiology, uses both "astronomic" and historical methods. Paleobiologists observe unique objects and relate them to general laws, but they also are very sensitive to the contingencies involved in linear historical development. Development in the individual human is also contingent, and developmental psychology has, at times, concerned itself with such "longitudinal" phenomena. What needs stressing more, however, is the way in which the unique individual is the cumulative product of a historical process. Thus, we often fail to see in the adult (or in the young child) the traces of processes and contingencies that occurred earlier in its history. The way in which the infant encounters and learns the world is not suddenly replaced by later functions or processes, but rather older individuals reveal the continuing influences and "remnants" of their "evolutionary" development.

There are, of course, large numbers of attempts at empirical and conceptual analyses in psychology and the other social senses that are nonexperimental. The best surviving example is probably psychoanalytic theory. Unfortunately, psychoanalytic theory fails the crucial test of cumulative accretion of empirical knowledge and theoretical consistency. Instead we have extensive internecine warfare with various "schools" claiming the mantle of succession to Freud or wanting to be seen as independent innovators. But there is no paradigmatic, established corpus. Nor have its original interdisciplinary goals been pursued in the succession to Freud.[73] No other attempt in psychology comes even close. In the other social sciences, there have been valiant attempts that failed for one reason or another: in sociology there was George Herbert Mead and Talcott Parsons, while in anthropology early attempts at systematization have now given way to a clearly postmodernist atomization and often trivialization of "the human science." In general, it must remembered that a true nonexperimental science such as astronomy and paleontology needs a long period of accumulation of data and theory—the cumulative accretion of common insights and common ways of viewing the world. That will take a long time—not just decades. Essentially the question about doing psychology without experiments is about psychology and the everyday world. Just as astronomers are concerned with everyday stars but supported by not very everyday theories, psychologists have begun to perceive how their theories may apply to the events of the real world.[74]

Envoi

One of the motivations for my presentation has been to demonstrate that there is no value-free psychology, that all of psychology is imbued with either explicit values or directed by implicit values that have their source in the social and historical context in which a particular psychological approach is situated. And in a wider sense our psychology reflects abiding traditions of Western society. Two such traditions that can also be seen in psychological conceptions of human nature have dominated Western thought. The deterministic and scientistic view sees humans flawed, often close to the Augustinian sense, whereas the free and humanist view requires a more optimistic and ameliorative view of human society. The constrained determinist view of humans tends toward a society that can deal with supposed human evil, and also needs a group that not only understands the human flaws but that can adequately control them. The humanist view, on the other hand, may leave open to individual choice matters that are often primarily social needs and concerns; i.e., it may be overly concerned with necessary social controls. But it surely can be argued that ideologies that bemoan the weak, sinful, and irresponsible character of humans are also the ones that see requirements for strong central control over those irrational tendencies and the curbing of popular sovereignty. Liberal democracy has still not come to a resolution between the problems of an unconstrained free market and a commitment to democratic controls.[75] An understanding by social scientists of these problems, of the consequences of constraint and of freedom, should help us resolve some of these problems, without—in the process—coming under the control of those who wish to determine the outcome of that resolution.

Notes

CHAPTER 1

1. As, for example, Burkert's (1996) masterly treatment of the pervasiveness and ubiquity of human adoptions of religious beliefs.
2. Gould, 1993, pp. 280–282; Tversky and Kahneman, 1973.
3. Gould, 1993, p. 281.
4. Sowell, 1986.
5. Sowell, 1986, p. 106.
6. Chapman and Jones, 1980.

CHAPTER 2

1. Marlowe, 1971.
2. For an extensive discussion, see Pagels, 1988.
3. Descartes, 1637/1960, p. 41.
4. Turing, 1950.
5. Descartes, 1637/1960, p. 42. But note in chapter 11 and in the Appendix how postmodern theories of mind postulate such "different devices."
6. Silverberg, 1980, p. 57.
7. Slater, 1959, p. 1046.
8. Lumsden and Wilson, 1983, p. 64.
9. Lumsden and Wilson, 1981, p. 36.
10. Lumsden and Wilson, 1981, p. 7.

11. J. M. Mandler, 1992.

12. Diamond and Carey, 1986.

13. G. A. Miller, 1956.

14. Piaget, 1954.

15. Baillargeon, Spelke, and Wasserman, 1985.

16. Baillargeon, Spelke, and Wasserman, 1985, p. 206.

17. Silverberg, 1978, 1980.

18. Silverberg, 1980.

CHAPTER 3

1. I am indebted to Luria, Gould and Singer (1981), whose book *A View of Life* has provided an excellent outline and approach for presenting some of this material. For an outstanding discussion of creationism, see Kitcher (1982).

2. Hamilton, 1964.

3. Dawkins, 1976.

4. See de Waal, 1996, pp. 13–20.

5. See Kropotkin, 1902.

6. Barash, 1980, p. 212.

7. See, for example, Wilson and Sober, 1994.

8. Gould and Vrba, 1982.

9. In this section, I have made use of the information provided elegantly by Gould (1991, pp. 139–151).

10. Mivart, 1871.

11. Kingsolver and Koehl, 1985.

12. Marden and Kramer, 1994.

13. Fischman, 1995, p. 365.

14. See, for example, Eldredge and Tattersall, 1982.

15. Casts of the inside of the brain cavity. Tobias, 1990, 1995.

16. Tobias, 1990. For an extensive review of related evidence see Tobias (1995). See also chapter 11.

17. Lieberman, 1989.

18. See, for example, Knecht et al., 1993.

19. Mayr, 1982.

20. For a reasonable argument that Neanderthals were a separate and dead-end development, see Stringer and Gamble (1993).

21. See Stringer and Gamble, 1993.

22. It is useful to be reminded that culture, defined as transmission of skills or knowledge across generations, is also not unique to humans (Mayr, 1982, p. 622).

23. Taken from Eldredge and Tattersall (1982).

24. Even these "universal" traits are likely to be variable (i.e., probabilistic) to some extent.

25. See, for example, Gould, 1980.

26. For the best extensive critique of sociobiology see Kitcher (1985); for two collections that represent both sides, see Gregory, Silver, and Sutch (1978) and Barlow and Silverberg (1980).

27. Washburn, 1978, p. 60.

28. Washburn, 1978, p. 60.

29. Beach, 1978, p. 131.

30. Beach, 1978, p. 131, italics in original.
31. Barlow and Silverberg, 1980, p. 14.
32. Barlow and Silverberg, 1980, p. 15.
33. Barash, 1980, p. 212.
34. Kitcher, 1985, 1990; Gould, 1977, p. 188ff.
35. This assumes for the sake of exposition that there are no prenatal differences in environments and no influences of such differences on genetic expression.
36. See, for example, Kamin, 1974.
37. It just does not seem to be reasonable to consider adoptive environments as "uncorrelated" (Plomin, 1990, p. 41).
38. According to the genetic argument they also behave alike, which affects their treatment, and the problem of interaction escalates.
39. See Berscheid's (1996) discussion of these issues, and her conclusion: "many morphological features . . . lead to inferences about other, less easily discerned, dispositional characteristics . . . (e.g., intelligence, personality)" (p. 5).
40. I discuss specific examples with respect to intelligence in chapter 9.
41. Plomin et al., 1990.
42. Prescott, Johnson, and McArdle, 1991, p. 376.
43. Plomin et al., 1990, p. 376.
44. McGue and Lykken, 1992, p. 368.
45. I am grateful to Ellen Berscheid for this example.
46. Davis, Phelps, and Bracha, 1995.
47. After all this work, one might still have to be aware of very specific interactions (e.g., how a particular genetic constitution interacts with specific environments). Consider, for example, a study of difference in male and female cognitive achievements in patriarchal versus matriarchal cultures.
48. I am indebted to Steve Jones' BBC Reith lecture for this observation.

CHAPTER 4

1. Deutsch, 1951, p. 216.
2. Searle, 1992, p. 18.
3. Lycan, 1987, p. 44.
4. The construction of "feelings" in particular demonstrates that the product is in part a function of cultural and social conditions incorporated into the constructive process. For an example of such cultural differences in the reaction to pain, see Zborowski (1969).
5. Turing, 1950.
6. See Dennett, 1991; Jackendoff, 1987.
7. Kant, 1781/1929, p. 183.
8. For an extensive discussion of the psychological use of schemas, see Rumelhart and Ortony (1978). For a more general discussion, see J. M. Mandler (1984b).
9. The reaction to expectations not met and to change are discussed at length in chapters 6 and 7.

CHAPTER 5

1. For more extensive and technical discussions of consciousness, see G. Mandler (1975a, 1984a,b, 1985, 1988, 1989, 1992c, 1996, in press a), from which parts of this chapter are taken.

2. For a discussion of that history, see G. Mandler (1985, chpt. 1) and Shallice (1991). See the Appendix for the more general social context.

3. Miller, 1985.

4. Thagard, 1986, p. 313.

5. For example, Jackendoff, 1987; Thagard, 1986.

6. Burks, 1984.

7. I am indebted to Robert van Gulick for pointing out this is the obvious outcome of Searle's (1983) position on intentionality.

8. Marcel, 1983a.

9. There are cases where intense stimuli preempt conscious contents. Thus, intense noises, lights, or pains will frequently become "automatically" conscious. However, such instances are in a sense also examples of the dominance of what is "important" in the environment.

10. See Marcel (1983a,b) on the rejection of the identity notion.

11. J.M. Mandler, 1984a.

12. See Schacter (1987) for a review.

13. See, for example, Reber (1996) for the yea-sayers and Shanks et al. (1994) for the nay-sayers.

14. I return to this important mechanism in greater detail in the discussion of the feedback function of consciousness.

15. Miller, 1956. For a review and extensive additional evidence, see Mandler and Shebo (1982). For cross-cultural evidence see Cole, Gay, and Glick (1968).

16. I have discussed elsewhere (G. Mandler, 1975b) the manner in which these momentary conscious states construct the phenomenal continuity and flow of consciousness.

17. We might note that connectionist models (e.g., Rumelhart and McClelland, 1985) apparently do not have any mechanisms for selecting what is attended, though they can account for the accentuation and dominance of events once they are attended to. They might also consider the need for a mechanism akin to consciousness as a limiting serial process.

18. With apologies to Susan Sontag's *The Volcano Lover.*

19. e.g., Gregory, 1981.

20. Jackendoff, 1987, p. 26; emphasis added.

21. Note that these activations are in addition to the usual flow of activation that takes place during unconscious processing (see, e.g., McClelland and Rumelhart 1981).

22. Posner and Snyder's (1975) hypothesis that conscious states preempt pathways by the inhibition of competing possibilities may in part be related to the assumption that such pathways are more available because of preferential additional activation.

23. For a summary see G. Mandler (1989).

24. G. Mandler, 1993b, in press a.

25. The "troubleshooting" function of consciousness is an important instance of such constructions.

26. The distinction between attention and consciousness has been made (Kahneman and Treisman, 1984; G. Mandler, 1985), and for the present purposes it will suffice to agree that attentional processes (under some definitions) will produce conscious contents, but that a conscious content does not presuppose prior attention.

27. Nisbett and Wilson, 1977.

28. See Lackner (1988) for this and many other examples.

29. Nielsen, 1963.

30. I am grateful to Roy D'Andrade for extensive discussions that permitted the development of these notions.

31. G. Mandler, 1975b.

32. For example, Joliot et al., 1994.

33. Crick and Mitchison, 1983, p. 112. REM sleep refers to a stage of sleep accompanied by rapid eye movements which has been shown to be frequently the occasion for dreams.

34. Hobson et al., 1987.

35. Freud, 1900/1975.

36. Hobson, 1988, p. 291.

37. For a further pursuit of the analogy between dreaming and creativity see G. Mandler (1995).

38. Crick and Mitchison, 1983; Robert, 1886.

39. See, for example, Weiskrantz, 1985.

40. Tulving, 1985.

41. Muenzinger, 1938.

42. Gould, 1993, p. 321.

43. Gleitman, Gleitman, and Shipley, 1972.

44. J. M. Mandler, 1988.

45. J. M. Mandler, 1984a.

46. Pippard, 1985, p. 9.

47. Putnam, 1980.

48. Such as Dennett's, 1991.

CHAPTER 6

1. An extensive discussion of my view on emotion can be found in two books (G. Mandler 1975b, 1984b) and in subsequent chapters (G. Mandler 1979a, 1988, 1990a,b, 1992a,b, in press b) from which some of this material is taken.

2. See Geertz, 1973.

3. The conviction with which these views are held may well be, in John Ruskin's felicitous phrase, examples of pathetic fallacies: "All violent feelings . . . produce in us a falseness in all our impressions of external things, which I would generally characterize as the 'Pathetic Fallacy'" (1843/1906, vol. iii, p. 148).

4. McNaughton, 1989, p. 3.

5. James, 1884, 1890, 1894.

6. James, 1894.

7. For example, Ortony, Clore, and Collins, 1988.

8. James, 1894, p. 193.

9. James, 1894, p. 194. I note, though, that the same bodily experiences have been known to accompany passionate love, extreme guilt, and other emotions.

10. Fridlund, 1991; G. Mandler, 1975b, 1984b; Ortony and Turner, 1990.

11. Geertz, 1973.

12. Cannon, 1929; Schachter and Singer, 1962.

13. e.g., Shallice, 1988.

14. Dewey, 1894; James, 1884; Paulhan, 1887.

15. This is the position initially taken by Schachter and Singer (1962) and which I have adopted.

16. As I have noted, my approach to emotion took off from Stanley Schachter's early work (Schachter and Singer, 1962, Schachter, 1970). I added notions about some of the origins of autonomic/visceral arousal in 1964, realizing later that they recapitulated in part the approaches of Paulhan (1887) and of Dewey (1894). Since then several

other writers have adopted these views of interruption and discrepancy (e.g., Oatley, 1992).

17. See, for example, Friedman, 1979; Loftus and Mackworth, 1978; J. M. Mandler and Johnson, 1976.

18. Cannon (1930).

19. There are two possible mechanisms for the registration of change. One is the usual perceptual one, which requires specific comparisons between expected and actual stimulation; the other makes use of the fact that the perceptual system can react to global changes without having yet analyzed the information as to "what" has changed. Whichever one is used, the alert from the SNS would still, given the slowness of the autonomic nervous system, be available later than the perceptual analysis.

20. For more extensive discussion of these issues, see G. Mandler (1984b).

21. Mandler, 1964, 1984b, 1990b; MacDowell and Mandler, 1989.

22. Ortony, Clore, and Collins, 1988.

23. Though I must emphasize that this is *not* the only mechanism for the production of SNS arousal.

24. See Berscheid (1982, 1983, 1991) for related and ingenious analyses of close relationships.

25. For a sensitive discussion of such interruptions, losses and emotions, see Berscheid (1984, 1991).

26. See Lutz and Abu-Lughod, 1990.

27. Abu-Lughod and Lutz, 1990, p. 11.

28. See, for example, Lutz, 1988.

29. Lutz, 1983, 1988.

30. Zajonc, 1980, 1984.

31. G. Mandler and Shebo, 1983.

32. Izard, 1972, p. 51.

33. Izard, 1972, p. 52.

34. Izard, 1972, p. 11.

35. Izard and Buechler, 1980, p. 169.

36. Frijda, 1986, e.g., p. 473.

37. Ortony, Clore, and Collins, 1988, p. 13.

38. Oatley and Johnson-Laird, 1987, p. 35.

39. Lazarus, 1991; Folkman and Lazarus, 1990.

40. Qtd. in Fridlund, 1992b, p. 119.

41. For an excellent presentation of modern views on facial expression, see Fridlund (1994), which also presents some of the arguments referenced to Fridlund in this section. For a general overview of contemporary methods and theories, see Russell and Fernández-Dols (1997).

42. Tomkins, 1962/1992, 1981.

43. Ekman, e.g., 1982.

44. Fridlund, 1991, 1992a; G. Mandler, 1975b, 1984b, 1992a.

45. Fridlund, 1991.

46. Bavelas et al., 1986.

47. Ekman, 1989. For an insightful critical review of the literature on cross-cultural consistencies in emotional displays, see Russell (1994).

48. Averill, 1980; Fridlund, 1991, 1992a ; Ortony and Turner, 1990.

49. Izard, 1977; Plutchik, 1980.

50. See G. Mandler, 1984b, p. 36.

51. For a survey of this contentious issue, see Ortony and Turner's original paper

(1990) and the rejoinders (Ekman, 1992; Izard, 1992; Panksepp, 1992; Turner and Ortony, 1992)

52. For example, Toda, 1982.

53. Wald, 1978, p. 277.

54. The fear of our animal ancestry also ignores the peacemaking proclivities of some of our primate cousins, explored in my discussion of aggression in chapter 8.

55. Kitcher, 1990.

56. Pick, 1970.

57. For example, Bem, 1967.

CHAPTER 7

Most of the material in this chapter has been adapted from a chapter on value (G. Mandler, 1993a, see also 1982). In addition, the Appendix is intended to illustrate the influence of social and cultural values on psychological thought.

1. Köhler, 1944, p. 199.

2. Köhler, 1944, p. 199.

3. The avoidance of something as vague as values was a symptom of the "physics envy" discussed in the Appendix.

4. For example, de Waal, 1996; Geertz, 1973; Schweder, 1982; Strum, 1987.

5. Ralph Barton Perry, 1926.

6. But note in the Appendix some possible pitfalls of the experimental method.

7. More technically, these are instrumental values represented by actions and terminal values represented by end-states.

8. Rokeach, 1973, p. 25

9. Rokeach, 1973, p. 23.

10. Köhler, 1938.

11. Köhler's presentation anticipates a later development that arose out of Gestalt psychology—the ecological psychology of James Gibson, whose "affordances" are similar to the requiredness invoked by Köhler.

12. Köhler, 1944, p. 204.

13. For a cognate discussion of the relationship between emotion and value, see Ortony (1991).

14. This approach is similar to Bem's descriptions of self-perception (1967).

15. Facial displays fall into this category. As I have noted in chapter 6, they are evolutionary remnants of a primitive nonverbal communication system that expresses value and informs the quality of emotions.

16. For a cognate approach, see Zajonc (1968).

17. See, for example, Brickman et al. (1972) and G. Mandler and Shebo (1983) for increasing negative value and Gaver and Mandler (1987) for positive changes in the case of music appreciation.

18. G. Mandler and Shebo, 1983.

19. See J. M. Mandler, 1984b, Rosch, 1978.

20. Purcell (unpublished data) has shown that when people judge the typicality (goodness of example), interestingness, and liking (preference) for a wide variety of buildings, one finds that typicality and preference are highly correlated ($r = .77$), particularly when interestingness was held constant ($r_{partial} = .94$). Typicality and interestingness, on the other hand, are negatively correlated ($r = -.29$, with preference controlled $r_{partial} = -.87$). Interestingness may be thought of as an index of deviation from typicality or of complexity. In this case, preference was directly related to typicality, independent of in-

terestingness. In another study of 20 selected paintings, Purcell again found a high correlation (.80) between the mean ratings of typicality and preference.

21. Van Orden and Uyeda, unpublished data.

22. It should be noted that the value aspect of "familiarity" is just one facet of the effect of prior experiences. Other aspects play an important role in human memory, where the symptom of "familiarity" is a central phenomenon in the automatic (implicit), contrasted with deliberate (explicit) access to mental contents (G. Mandler, 1989, 1994b).

23. Arnheim, 1966, p. 125.

24. G. Mandler, 1984b.

25. Arnheim, 1966, p. 124.

26. The attraction of play might have similar roots.

27. For example, the creative individual who seeks out complexity (see Dellas and Gaier, 1970).

28. Quoted in Rosen, 1971, p. 393.

29. For more extensive discussions of this and related issues, see G. Mandler (1984b, 1995).

30. Berger, 1988, p. 33.

31. Meyer, 1956.

32. Rosen, 1971, pp. 58, 74, 83.

33. The influence of a culture and its values on all its products is one of the themes of my discussion of psychology's recent history in the Appendix.

34. See, for example, the literature on partial reinforcement and extinction.

35. It is sometimes overlooked that even Marx was a gradualist in this sense; he saw changes in social consciousness developing slowly after the proletarian revolution (cf., for example, his discussion in the *Critique of the Gotha Programme*).

36. See P. Mandler, 1989.

CHAPTER 8

1. Gottlieb, 1984, p. 94.

2. Gottlieb, 1984, pp. 116–117.

3. Gumplowicz, 1885/1963, p. 40.; emphasis added and translation improved.

4. For an extensive discussion of human aggression, including the division among types of aggression, see Fromm (1973).

5. The distinction has sometimes been referred to as one between instrumental and expressive aggression.

6. See, for example, Lorenz, 1963.

7. Other versions of the drive to aggression are found in Freud (1916/1975) and in Dollard et al. (1939).

8. Much of this argument is heavily indebted to Strum (1987).

9. de Waal, 1989, p. 270. I return to de Waal and his work on animal morality in chapter 10.

10. It is interesting to note that under this model, aggressive behavior can also be acquired in relation to an obstructing nonsocial environment.

11. See, for example, Staub (1996) for a discussion of these conditions.

12. For an extensive and instructive analysis of aggression from the point of view of social learning theory, see Bandura (1973).

13. Most of this discussion is taken from Dentan (1968).

14. For description for another relatively nonaggressive society that denies any feeling of anger, see Briggs's (1970) description of the Utku, an Inuit tribe.

15. The information described below is taken from John Nance's book (1975), as well as from briefer descriptions by Elizalde (1971), Fernandez and Lynch (1972), and reports in Yen and Nance (1976).

16. Fox, 1976; Nance, 1975.

17. See, for example, Molony and Tuan, 1976.

18. Yen, 1976, pp. 182–183.

19. Nance, 1975, p. 444.

20. Lumsden and Wilson, 1983, p. 149.

21. Lumsden and Wilson, 1983, p. 160.

22. Darwin, 1871.

23. Lumsden and Wilson, 1983, p. 160.

24. Lumsden and Wilson, 1983, p. 160.

25. Chagnon, 1980.

26. There is some suggestion that different Yanonamo tribes may differ quite strongly in their aggressiveness. Reports from Brazil also suggest, in addition, that with increasing contact with the surrounding destructive and aggressive civilization which represents a real threat to the Yanomamos' survival, they have become less aggressive.

27. Thwaites, 1906.

28. Leacock, 1980.

CHAPTER 9

1. See, for example, the work of Sternberg, 1985, and Gardner, 1983.

2. For an extensive and consensual (and therefore not very satisfying) summary of most aspects of the meaning and uses of the IQ, see Neisser et al. (1996).

3. Jensen, 1978, Lewontin, 1970, Mackenzie, 1980.

4. See also Gould, 1981.

5. Bouchard et al., 1990.

6. Bouchard et al., 1990, Table 3.

7. Flynn, 1987.

8. Flynn, 1987.

9. These data are taken from Flynn (1994.)

10. Neisser et al., 1996.

11. Flynn, 1994.

12. It should be noted that less than 100 years ago, similar differences in IQ were noted between immigrant (e.g., Jewish and Irish) and native American populations (and equally assigned to genetic causes).

13. Ogbu, 1978, 1994.

14. Ogbu, 1978, p. 213.

15. Ogbu, 1978, p. 214.

16. Quoted in Ogbu, 1978, p. 309.

17. Quoted in Ogbu, 1978, p. 283.

18. Ogbu, 1978.

19. See Lavin and Hyllegard (1996) for a description and some of the surprising results of the open admissions policy instituted in New York city colleges in 1970.

20. Ford and Beach, 1952.

21. Heider, 1976.

22. Maccoby and Jacklin, 1974; Halpern, 1986 (whose overview and review I follow here).

23. Leacock, 1980; Barash, 1977.

24. Barash, 1977, pp. 301–302.

25. See Halpern (1986) for a more detailed discussion.

26. Most of the following material is taken from Bareh (1985).

27. Ahmed, 1994.

28. Speculations about genetic aspects of complex sexual and mating behaviors that abounded in the early days of sociobiology have slowly given way to more empirical investigations. But even the most ambitious attempts (e.g., Buss, 1989) still fall short in the appropriate sampling method and considerations of alternate hypotheses and models.

CHAPTER 10

Much of the contents of this chapter is based on two previous expositions (G. Mandler, 1993b, 1994a).

1. The principles of a moral life have a variety of sources, such as exemplars to be imitated and followed, that eventually come to be enshrined in rules and principles.

2. I am indebted to a reading of Elaine Pagels' *Adam, Eve, and the Serpent* (1988) for this exposition.

3. This is echoed in contemporary life when religious and political fundamentalists decry government interference but welcome the state's statutory opposition to abortion, regulation of homosexuality, etc.

4. de Waal, 1996, p. 211. See de Waal (1996) for a highly accessible and rich description of chimpanzees and other primates and insightful discussions of the literature and relevant evolutionary theory.

5. Goldhagen, 1996; (emphasis added). Goldhagen describes in detail both the origin of this moral incorporation and the manner in which it was carried out by some of the average German citizens who participated in its execution.

6. MacIntyre, 1981, p. 11.

7. James, 1890, vol. 1, p. 677.

8. Nor does it help to define (as some philosophers have) practical reason as "what it makes sense to do." Whose sense?

9. By adopting such a pragmatic, behavioral approach to objectivism and relativism, I avoid philosophical distinctions among such arcane topics as moral realism, normative relativity, constructivism, etc.

10. MacIntyre, 1988.

11. Nagel, 1988.

12. I remember an interchange between a philosopher and psychologist in which the former condemned a particular psychological theoretical model of human thought as producing only "pseudo-rationality." The psychologist replied: "But that is all there is."

13. For example, Hall, 1985; Mann, 1986.

14. Chomsky, 1991.

15. Berlin, 1969.

16. Partridge, 1967, p. 222.

17. Berlin, 1969, p. xxxix.

18. See Oppenheim, 1968.

19. Berlin, 1969.

20. Oppenheim, 1968, p. 558.

21. See Dworkin (1991) for a relevant argument and a practical application of Berlin's distinction.

22. The quite different issue of determinism and free will has been addressed previously in part (G. Mandler and Kessen, 1974).

23. Kierkegaard, 1844/1957.

24. In contrast, as I have indicated earlier, nations that have previously enjoyed democratic liberties generate particularly strong emotional reactions to repression and its removal; they "know" what is constrained. Conversely, it would be of interest to investigate formally the proposition that in the absence of repression, societies that have always enjoyed unrestricted liberties value these freedoms less.

25. Berscheid, 1983.

26. I find this another difficult case for the proponents of a limited number of basic emotions. It is, like love and lust, never mentioned in lists of such emotions.

27. I return to the issue of the meaning of power later; in the present context, it refers to the fact that the individual shares some real or imaginary ability to constrain others and to be free from constraints themselves.

28. Mann, 1986, p. 6.

29. For present purposes I do not employ the otherwise useful distinction among economic, ideological, and political power (see Hall, 1985; Mann, 1986).

30. Partridge, 1967, p. 224.

31. Berlin, 1969, p.xlv.

32. Freud's notion of "identification with the aggressor" exemplifies this phenomenon in part.

33. Hall, 1985.

34. In an analysis of nationalism in Eastern Europe, Gellner (1991) has noted that with modernity and industrialism a new context-free "High Culture" became the style of the entire society, and citizenship required its mastery. If one could not "assimilate into the dominant High Culture" then one needed "to ensure that one's own [national] culture becomes the defining one . . ." (Gellner, 1991, p. 130). It is the identification with such a national culture, which is powerless or unattainable or economically impotent in the context of the "High Culture" (e.g., Eastern European/Soviet culture), that provides some of the psychological background for extreme nationalism.

35. It should be noted that some of the leaders of the French revolution, following Rousseau, were advocating constructed liberties as well, thus producing another case of compounded natural and constructed liberties.

36. The fact that "pursuit of happiness" was substituted for John Locke's "property" in an early draft of the American Declaration of Independence illustrates how changes in the social definition of rights and liberties may overcome the attempt by empowered groups to enshrine their privileges as "rights."

CHAPTER II

Some of the material in this chapter is taken from my book on cognitive psychology (G. Mandler, 1985).

1. See G. Mandler (1994b) for a summary of these trends as well as for a reminder of the fragility and lack of accretion in psychological research. In particular, I point to the oft-forgotten insight by Katona (1940) and extensive research in the 1960s and 1970s that the organization of memorial material is a necessary and sufficient condition for its retrieval.

2. For some interesting speculations about the evolution about the human mind, see Donald (1991).

3. G. Mandler, 1979b.

4. See J. M. Mandler, 1984b.

5. See Pinker (1994) for ex cathedra pronouncements and Pinker and Bloom (1990

and commentaries by many others) for the astounding variety that can be found in the field.

6. Pinker, 1994.

7. Chomsky, 1972.

8. For example, Bates et al., 1989.

9. Wundt, 1900, vol. 1, part II, p. 606.

10. Wundt, 1896, p. 350.

11. Wundt, 1896, p. 350.

12. Wundt, 1900, vol. 1, part I.

13. Andrews and Stringer, 1993.

14. Givòn, 1979.

15. Pinker, 1994.

16. Gould, 1993, p. 321.

17. See the discussion of the work of Tobias in chapter 3.

18. For example, current basic disagreements between cognitive and transformational linguists suggest different evolutionary histories.

19. Chomsky, 1972; Pinker, 1994.

20. See, for example, Gentner and Stevens, 1982.

21. J.M. Mandler, 1984b.

22. Which, in turn, is different from the underlying representation that generates the mental model.

23. For older critiques see G. Mandler, 1969, 1985.

24. Pippard, 1985, p. 9.

25. Putnam, 1980, p. 142. A similar view of the reductionist fallacy argues that all phenomena have a finite distribution in nature, which can range from a very broad one (in the case of the physicists' particles, which range over all matter) to very narrow ones (in the case of human social organization). The misleading reductionist claim is that the narrowly distributed properties can be explained in terms of the more broadly distributed ones; i.e., mind can be explained in terms of matter (Eldredge and Tattersall, 1982).

26. Pinker, 1994.

APPENDIX

This appendix is a revised version of a paper originally published as "The situation of psychology: Landmarks and choicepoints" in *American Journal of Psychology* (1996, vol. 109, pp. 1–35). I am most grateful to a number of friends and colleagues who read and commented on previous versions of this chapter. In particular, comments by Patricia Kitcher and Tim Shallice significantly influenced some sections, and I also benefited from comments by Kimberly Jameson, William Kessen, Peter Mandler, and the members of the Science Studies group at the Department of Psychology, University College London.

1. The scope of "psychology" discussed here is mainstream theoretical psychology, often restricted to being "experimental" (currently "cognitive"), but generally excluding specific concerns of social, clinical, or personality psychologies, though these are not untouched by the topics I discuss.

2. The discussion of social influences on scientific labors—the situating of science as culture—has become of interest to philosophers, sociologists, and historians of science. The movement started in a serious way in the 1930s. See, for example, Fleck's (1935/1979) tour de force on the treatment of syphilis and Marxist analyses initially marked by rather extravagant treatments such as that of Newton by Hessen (1931/1971). For more contemporary approaches, see Henderson (1990) and Pickering (1984). It has now become a

general enterprise, and for a refreshing antidote to some of its excesses, see the balanced view of Kitcher (1993). In psychology, this kind of approach has been relatively rare. For an interesting relevant selection, see Buss (1979).

3. I share with others serious doubts about the generality of the concept of "progress." Human history and endeavor is replete with change, but whether any such change is seen to be progress depends on the set of values within which one wants to interpret such a change.

4. These trends were probably part of a more general nineteenth century trend of seeking historical explanations for complex phenomena.

5. Much of this and the following information is taken from Wundt's autobiography, *Erlebtes und Erkanntes* (1920).

6. Sheehan, 1978.

7. Wundt, 1920, p. 195.

8. Wundt, p. 197.

9. The term *Geisteswissenschaften* was introduced with the translation of J. S. Mill's *System of Logic* in 1849 and Mill's "moral sciences" (Makkreel and Rodi, 1989).

10. Makkreel and Rodi, 1989, especially pp. 12–13.

11. Dilthey, 1883/1959, p. 11.

12. Wundt, 1920, p. 201. I discuss more about the *Völkerpsychologie*, the 10-volume ethnopsychology, below. It should also be noted that, in parallel with his psychological and physiological interests, Wundt continued to write on philosophical matters, logic and ethics in particular.

13. Wundt, 1920, p. 301

14. Wundt's discussion in his autobiography (1920) of the problem of finding more space for his institute has a very contemporary academic flavor.

15. Wundt, 1920, pp. 380–381.

16. See, for example, Schnabel, 1950.

17. See, for example, Stern, 1961.

18. Porter, 1986.

19. See, for example, Blumenthal's, 1970, discussion.

20. I am grateful to Patricia Kitcher for suggesting this line of argument.

21. See the discussion in chapters 4 and 11.

22. See, for example, Sheehan, 1978, pp. 234–235.

23. Tuchman, 1966, p. 307.

24. For other dimensions of Wundt's psychology, often at variance with the conventional view, see the important contribution by Blumenthal (1970).

25. See Porter, 1986, p. 67.

26. The best translation for the title of *Völkerpsychologie* is probably "ethnopsychology," since Wundt held the topics that he considered there to be cultural products.

27. Wundt, 1900–1909, vol. 1, part 1, introduction.

28. Wundt, 1912, 1916.

29. Wundt, 1916.

30. Thagard, 1986.

31. In this account, I have had little occasion to do justice to the British empiricists. I should, however, note the last great one who provided a bridge to modern psychology, Alexander Bain. Bain thought deeply on a variety of theoretical issues, constrained by a commitment to associationism. He apparently originated the notion of representation, when he talked of experiences—the original states of consciousness—as primary states, and their revived states as "representations" (Bain, 1855). And I will always treasure a marginal remark made by a former owner of my copy of Bain's discussion of consciousness.

In 1891, a Grace A. Brubaker wrote in the margin: "This is a long lesson and little am I the wiser."

32. These functionalists were the forerunners of contemporary evolutionary psychology, as illustrated by some parts of this book.

33. Wundt's "introspection" was devoted to self-observation, producing the protocol sentences of psychology.

34. One of the intellectual (and personal) tragedies of late nineteenth-century psychology was the fact that the work of James Mark Baldwin (e.g., 1894) never had as a large an impact as it should have, in part because he was hounded out of the academy for of a personal episode distasteful to Victorian mores. Some of Baldwin's work was an important precursor to Piaget's contribution to mental development.

35. See Tuchman's (1966) accessible presentation of the tensions and history of that period.

36. Hughes, 1958.

37. Hughes, 1958, p. 4.

38. See Mandler and Mandler (1964) for a discussion of some of those developments.

39. That tradition of going from thought/behavior directly to physiology was continued in the United States by the radical behaviorists, specifically by B. F. Skinner, in the 1930s and by radical philosophers in the 1980s.

40. See, for example, von Hartmann (1869) for an earlier extensive discussion of the unconscious.

41. See Mandler and Mandler, 1964.

42. There was a continuing concern with the appropriate level or nature of explanation in psychology. In 1904 Henry Watt, in an influential paper abstracted in Watt (1905–06), noted that he had deliberately omitted any physiological explanation in order to "tempt" the physiologically sophisticated to try to undertake such an endeavor.

43. Frijda and deGroot, 1981.

44. Important was Selz's (1913) rejection of Bain's constellation theory, essentially an additive theory of associative strengths. To give a simplified example, Bain postulated that when required to give the opposite of "black," the associates of "black," and of "opposites" would be generated and when summed would result in the production of "white," which is an associate of both "black" and "opposite." But Selz noted that this explanation is not determinate enough, since "night" would also fulfill both conditions, but would not be given in response to the question. Instead of these explanations, Selz stressed the importance of relational concepts such as "opposite," "next to," etc. Selz was killed in a German concentration camp.

45. Tuchman, 1966, p. 161.

46. Wiebe, 1967.

47. Watson, 1913.

48. Watson, 1919, p. vii.

49. Esper, 1964, p. v.

50. Only Skinner strayed from the physicalist definitions of stimuli and responses and introduced a functionalist (nearly cognitive) definition of eliciting conditions and organisms' behavior.

51. In addition, the early 20th century saw the development of atomistic physiological associationism in Russia. Pavlovian ideas were soon to be absorbed by the behaviorist movement. But I wonder why and how such similarities developed in the two countries and what conditions and precursors they have in common.

52. Taylor, 1911. For other discussions of the parallel between the two movements, see Schwartz (1986).

53. For an account of the emigration to the United States and the resistance of the mainstream of the academic establishment, see Mandler and Mandler (1968).

54. These were useful in the early 1950s in a precursor movement to cognitive psychology.

55. Organization had previously been championed by Gestalt psychology (e.g., Katona, 1940).

56. It should be remembered that announcements of a "new" psychology have occurred several times during the past 200 years.

57. Bartlett, 1932; Broadbent, 1958; Craik, 1943, 1966.

58. Binet, 1894, 1903 (see also Pollack and Brenner, 1969); Claparède, 1934; Köhler, 1917, 1925, 1929.

59. After 1932 there was obviously no important psychology being done in Germany, and the postwar rehabilitation took too long to influence significantly the events of the 1950s.

60. See chapter 4.

61. Kant, 1781/1929; Bartlett, 1932; Piaget, 1953; Rumelhart and Ortony, 1978.

62. See, for example, the then-influential book by Snygg and Combs (1949). The marginal nature of cognition is exemplified in a curious outburst against cognitivist approaches by three Young Turks (J. Deese, J. J. Jenkins, G. Mandler) at a 1959 conference that marked a boundary between the old and the new (Cofer, 1961).

63. Bruce, 1994.

64. Piaget, 1954.

65. Significant changes in direction, in theory, and in the focus of experimentation occurred in memory (Cofer, 1961; Jenkins, 1955), attention (Broadbent, 1958; Treisman, 1964), emotion (Schachter and Singer, 1962), information and problem solving (e.g., Newell and Simon, 1956), perception (Hochberg, 1968), developmental psychology (in the rediscovery of Piaget), and, of course, in George Miller's seminal article on the limits of perception and conception (1956), and in the Miller, Galanter, and Pribram volume (1960). The same turmoil was discernible in cognate fields, such as artificial intelligence (Mechanisation of Thought Processes, 1959), anthropology (Goodenough, 1956; Lounsbury, 1956), and linguistics (Chomsky, 1956, 1957). For a more detailed discussion, see G. Mandler (1985, chpt. 1). For a comprehensive survey from an optimistic point of view, see Baars (1986).

66. Can it be a coincidence that the changes in psychology that concern me here all occurred about 20 to 25 years (a generation) later than the cognate major changes in the natural sciences and the humanities? Wundt was about 25 years later than changes in the natural sciences, both the behaviorist and the Gestalt reconstructions occurred some 25 years after the major intellectual changes of the 1890s, and the "cognitive revolution" was effective some 15 to 20 years after related changes in the sciences of information and communication.

67. See Dennett, 1991; Jackendoff, 1987.

68. Gergen, 1994.

69. The trend toward postmodernism was foreshadowed with the arrival of ethnomethodology in the 1960s.

70. See also chapter 11.

71. The term "physics envy" first appeared in the early 1970s to describe an attitude among biologists. It applies just as well, if not better, to psychology.

72. For example, Deese, 1985; Postman, 1988.

73. See Kitcher's (1992) important insights on this issue.

74. One outstanding recent example is in the application of Tversky and Kahneman's

work (1973) on prediction and representativeness to "real" politics, as seen in the important contributions of Sam Popkin to American politics (Popkin, 1991).

75. F. A. Hayek, in seeing that an unconstrained free market tends to lead to chance determining success rather than individual prowess or achievement, considered as a consequence the need to constrain democratic expression and to assert the rule of law free of democratic controls.

References

Abu-Lughod, L., & Lutz, C. (1990). Introduction: emotion, discourse, and the politics of everyday life. In C. Lutz & L. Abu-Lughod (Eds.), *Language and the politics of emotion.* Cambridge: Cambridge University Press.

Ahmed, S. Z. (1994, February 15). What do men want? *New York Times,* p. A19.

Andrews, P., & Stringer, C. (1993). The primates' progress. In S. J. Gould (Ed.), *The book of life* (pp. 219–251). New York: Ebury Hutchinson.

Arnheim, R. (1966). *Toward a psychology of art.* Berkeley, CA: University of California Press.

Averill, J. R. (1980). A constructivist view of emotion. In R. Plutchik & H. Kellerman (Eds.), *Theories of emotion.* New York: Academic Press.

Baars, B. J. (1986). *The cognitive revolution in psychology.* New York: Guilford Press.

Baillargeon, R., Spelke, E. S., & Wasserman, S. (1985). Object permanence in five-month-old infants. *Cognition, 20,* 191–208.

Bain, A. (1855). *The senses and the intellect.* London: J. W. Parker.

Baldwin, J. M. (1894). *Mental development in the child and the race, methods and processes.* New York: Macmillan.

Bandura, A. (1973). *Aggression: A social learning analysis.* Englewood Cliffs, NJ: Prentice-Hall.

Barash, D. P. (1977). *Sociobiology and behavior.* Amsterdam: Elsevier.

Barash, D. P. (1980). Predictive sociobiology: Mate selection in damselfishes and brood defense in white-crowned sparrows. In G. W. Barlow & J. Silverberg (Eds.), *Sociobiology: Beyond nature/nurture?* Washington, DC: American Association for the Advancement of Science.

Bareh, H. (1985). *The history and culture of the Khasi people*. Guwahati, India: Spectrum Publications.

Barlow, G. W., & Silverberg, J. (Eds.). (1980). *Sociobiology: Beyond nature/nurture?* Washington, DC: American Association for the Advancement of Science.

Bartlett, F. C. (1932). *Remembering*. Cambridge: Cambridge University Press.

Bates, E., Thal, D., & Marchman, V. (1991). Symbols and syntax: A Darwinian approach to language development. In N. Krasnegor, D. Rumbaugh, E. Schiefelbusch, & M. Studdert-Kennedy (Eds.), *Biological and behavioral determinants of language development* (pp. 29–65). Hillsdale, NJ: Lawrence Erlbaum Associates.

Bavelas, J. B., Black, A., Lemery, C. R., & Mullett, J. (1986). I show how you feel: Motor mimicry as a communicative act. *Journal of Personality and Social Psychology, 50,* 322–329.

Beach, F. A. (1978). Sociobiology and interspecific comparisons of behavior. In M. S. Gregory, A. Silver, & D. Sutch (Eds.), *Sociobiology and human nature* (pp. 116–125). San Francisco: Jossey-Bass.

Bem, D. J. (1967). Self-perception: An alternative interpretation of cognitive dissonance phenomena. *Psychological Review, 74,* 183–200.

Berger, J. (July 1988). The art of the interior. *New Statesman & Society, 1,* No. 6, 32–34.

Berlin, I. (1969). *Four essays on liberty*. London: Oxford University Press.

Berscheid, E. (1982). Attraction and emotion in interpersonal relationships. In M. S. Clark & S. T. Fiske (Eds.), *Affect and cognition: The Seventeenth Annual Carnegie Symposium on Cognition* (pp. 37–54). Hillsdale, NJ: Lawrence Erlbaum Associates.

Berscheid, E. (1983). Emotion. In H. H. Kelley, E. Berscheid, A. Christensen, J. H. Harvey, T. L. Huston, G. Levinger, E. McClintock, L. A. Peplau, & D. R. Peterson (Eds.), *Close relationships* (pp. 110–168). San Francisco: Freeman.

Berscheid, E. (1984). Emotional experience in close relationships: Implications for child development. In Z. Rubin & W. Hartup (Eds.), *The effects of early relationships on children's socioemotional development*. New York: Cambridge University Press.

Berscheid, E. (1991). The emotion-in-relationships model: Reflections and update. In W. Kessen, A. Ortony, & F. Craik (Eds.), *Memories, thoughts, emotions: Essays in honor of George Mandler* (pp. 323–335). Hillsdale, NJ: Lawrence Erlbaum Associates.

Berscheid, E. (1996). The "paradigm of family transcendence": Not a paradigm, questionably transcendent, but valuable nonetheless. *Journal of Marriage and the Family, 58,* 1–9.

Binet, A. (1894). *Introduction à la psychologie experimentale*. Paris: F. Alcan.

Binet, A. (1903). *L'étude experimentale de l'intelligence*. Paris: Schleicher Frères.

Blumenthal, A. L. (1970). *Language and psychology: Historical aspects of psycholinguistics*. New York: Wiley.

Bouchard, T. J., Jr., Lykken, D. T., McGue, M., Siegal, N. L., & Tellegen, A. (1990). Sources of human psychological differences: The Minnesota study of twins reared apart. *Science, 250,* 223–228.

Brickman, P., Redfield, J., Harrison, A. A., & Crandall, R. (1972). Drive and predisposition as factors in the attitudinal effects of mere exposure. *Journal of Experimental Social Psychology, 8,* 31–44.

Briggs, J. L. (1970). *Never in anger: Portrait of an Eskimo family*. Cambridge, MA: Harvard University Press.

Broadbent, D. E. (1958). *Perception and communication*. London: Pergamon Press.

Bruce, D. (1994). Lashley and the problem of serial order. *American Psychologist, 49,* 93–103.

Burkert, W. (1996). *Creation of the sacred: Tracks of biology in early religions*. Cambridge, MA: Harvard University Press.

Burks, A. (1984). *An architectural theory of human consciousness* (Technical Report No. 254). Ann Arbor: University of Michigan.

Buss, A. R. (1979). *Psychology in social context.* New York: Irvington.

Buss, D. M. (1989). Sex differences in human mate preferences: Evolutionary hypotheses tested in 37 cultures. *Behavioral and Brain Sciences, 12,* 1–49.

Cannon, W. B. (1929). *Bodily changes in pain, hunger, fear and rage.* New York: Appleton-Century-Crofts.

Cannon, W. B. (1930). The Linacre lecture on the autonomic nervous system: An interpretation. *Lancet, 218,* 1109–1115.

Chagnon, N. A. (1980). Kin-selection theory, kinship, marriage and fitness among the Yanomamo Indians. In G. W. Barlow & J. Silverberg (Eds.), *Sociobiology: Beyond nature/nurture?* Washington, DC: American Association for the Advancement of Science.

Chapman, A. J., & Jones, D. M. (1980). *Models of man.* Leicester: British Psychological Society.

Chomsky, N. (1956). Three models for the description of language. *IRE Transactions on Information Theory, IT-2(3),* 113–124.

Chomsky, N. (1957). *Syntactic structures.* The Hague: Mouton.

Chomsky, N. (1972). *Language and mind.* New York: Harcourt, Brace & World.

Chomsky, N. (1991). *Deterring democracy.* London: Verso.

Claparède, E. (1934). *La genèse de l'hypotheses.* Geneva: Kundig.

Cofer, C. N. (1961). *Verbal learning and verbal behavior.* New York: McGraw-Hill.

Cole, M., Gay, J., & Glick, J. (1968). A cross-cultural investigation of information processing. *International Journal of Psychology, 3,* 93–102.

Craik, K. J. W. (1943). *The nature of explanation.* Cambridge: Cambridge University Press.

Craik, K. J. W. (1966). *The nature of psychology.* (S. L. Sherwood, Ed.). Cambridge: Cambridge University Press.

Crick, F., & Mitchison, G. (1983). The function of dream sleep. *Nature, 304,* 111–114.

Darwin, C. (1871). *The descent of man, and selection in relation to sex.* New York: D. Appleton.

Darwin, C. (1872a). *The expression of the emotions in man and animals.* London: John Murray.

Darwin, C. (1872b). *The origin of species by means of natural selection, or the preservation of favoured races in the struggle for life* (6th ed.). London: John Murray.

Davis, J. O., Phelps, J. A., & Bracha, H. S. (1995). Prenatal development of monozygotic twins and concordance for schizophrenia. *Schizophrenia Bulletin, 21,* 357–366.

Dawkins, R. (1976). *The selfish gene.* Oxford: Oxford University Press.

Deese, J. (1985). *American freedom and the social sciences.* New York: Columbia University Press.

Dellas, M., & Gaier, E. L. (1970). Identification of creativity. *Psychological Bulletin, 73,* 55–73.

Dennett, D. C. (1991). *Consciousness explained.* Boston: Little, Brown & Company.

Dentan, R. K. (1968). *The Semai, a nonviolent people of Malaya.* New York: Holt, Rinehart and Winston.

Descartes, R. (1637/1960). *Discourse on Method and Mediations,* part V. Indianapolis: Bobbs-Merill.

Deutsch, K. W. (1951). Mechanism, teleology, and mind. *Philosophy and Phenomenological Research, 12,* 185–223.

de Waal, F. B. M. (1989). *Peacemaking among primates.* Cambridge, MA: Harvard University Press.

de Waal, F. B. M. (1996). *Good natured: The origins of right and wrong in humans and other animals.* Cambridge, MA: Harvard University Press.

Dewey, J. (1894). The theory of emotion I. Emotional attitudes. *Psychological Review, 1,* 553–569.

Diamond, R., & Carey, S. (1986). Why faces are and are not special: An effect of expertise. *Journal of Experimental Psychology: General, 115,* 107–117.

Dilthey, W. (1883/1959). *Einleitung in die Geisteswissenschaften.* Stuttgart: B. G. Teubner.

Dollard, J., Doob, L. W., Miller, N. E., Mowrer, O. H., & Sears, R. R. (1939). *Frustration and aggression.* New Haven, CT: Yale University Press.

Donald, M. (1991). *Origins of the modern mind: Three stages in the evolution of culture and cognition.* Cambridge, MA: Harvard University Press.

Dworkin, R. (1991). Liberty and pornography. *The New York Review of Books, 38*(14), 12–15.

Ekman, P. (Ed.). (1982). *Emotion in the human face.* Cambridge: Cambridge University Press.

Ekman, P. (1989). The argument and evidence about universals in facial expressions of emotion. In H. Wagner & A. Manstead (Eds.), *Handbook of Social Psychophysiology* (pp. 143–164). Chichester, England: John Wiley & Sons.

Ekman, P. (1992). Are there basic emotions? *Psychological Review, 99,* 550–553.

Eldredge, N., & Tattersall, I. (1982). *The myths of human evolution.* New York: Columbia University Press.

Elizalde, M. J. (1971). *The Tasaday forest people.* Cambridge, MA: Smithsonian Institution, Center for Short-Lived Phenomena.

Esper, E. A. (1964). *A history of psychology.* Philadelphia: W. B. Saunders.

Fernandez, C. A., II, & Lynch, F. S. J. (1972). The Tasaday: Cave-dwelling food gatherers of South Cotabato, Mindanao. *Philippine Sociological Review, 20,* 279–330.

Fischman, J. (1995). Arms and the man. *Science, 268,* 364–365.

Fleck, L. (1935/1979). *Genesis and development of a scientific fact.* (T. J. Trenn & R. K. Merton, Eds., F. Bradley & T. J. Trenn, Trans.). Chicago: University of Chicago Press. (Originally work published by Schwabe, Basel, 1935).

Flynn, J. R. (1987). Massive IQ gains in 14 nations: What IQ tests really measure. *Psychological Bulletin, 101,* 171–191.

Flynn, J. R. (1994). IQ gains over time. In R. J. Sternberg (Ed.), *Encyclopedia of human intelligence.* New York: Macmillan.

Folkman, S., & Lazarus, R. S. (1990). Coping and emotion. In N. S. Stein, B. L. Leventhal, & T. Trabasso (Eds.), *Psychological and biological approaches to emotion* (pp. 313–332). Hillsdale, NJ: Lawrence Erlbaum Associates.

Ford, C. S., & Beach, F. A. (1952). *Patterns of sexual behavior.* New York: Harper.

Fox, R. B. (1976). Notes on the stone tools of the Tasaday, gathering economies in the Philippines, and the archaeological record. In D. E. Yen & J. Nance (Eds.), *Further studies on the Tasaday.* Makati, Philippines: Panamin Foundation.

Freud, S. (1900/1975). The interpretation of dreams. In *The Standard Edition of the Complete Psychological Works of Sigmund Freud.* London: Hogarth Press.

Freud, S. (1916/1975). Introductory lectures on psychoanalysis. In *The Standard Edition of the Complete Psychological Works of Sigmund Freud.* London: Hogarth Press.

Fridlund, A. J. (1991). Evolution and facial action in reflex, social motive, and paralanguage. *Biological Psychology, 32,* 3–100.

Fridlund, A. J. (1992a). The behavioral ecology and sociality of human faces. In M. S. Clark (Ed.), *Review of Personality and Social Psychology* (pp. 90–121). Beverly Hills, CA: Sage.

Fridlund, A. J. (1992b). Darwin's anti-Darwinism in the expression of the emotions in man and animals. In K. T. Strongman (Ed.), *International Review of Studies on Emotion* (pp. 117–137). Chichester, England: John Wiley & Sons.

Fridlund, A. J. (1994). *Human facial expression: An evolutionary view.* San Diego, CA: Academic Press.

Friedman, A. (1979). Framing pictures: The role of knowledge in automatized encoding and memory for gist. *Journal of Experimental Psychology: General, 108,* 316–355.

Frijda, N. H. (1986). *The emotions.* Cambridge: Cambridge University Press.

Frijda, N. H., & deGroot, A. D. (1981). *Otto Selz: His contribution to psychology.* The Hague: Mouton.

Fromm, E. (1973). *The anatomy of human destructiveness.* New York: Holt, Rinehart and Winston.

Gardner, H. (1983). *Frames of mind: the theory of multiple intelligences.* New York: Basic Books.

Gaver, W., & Mandler, G. (1987). Play it again, Sam: On liking music. *Cognition and Emotion, 1,* 259–282.

Geertz, C. (1973). *The interpretation of cultures: Selected essays.* New York: Basic Books.

Gellner, E. (1991 September–October). Nationalism and politics in Eastern Europe. *New Left Review, 127–134.*

Gentner, D., & Stevens, A. (1982). *Mental models.* Hillsdale, NJ: Lawrence Erlbaum Associates.

Gergen, K. J. (1994). Exploring the postmodern: Perils or potentials? *American Psychologist, 49,* 412–416.

Givòn, T. (1979). *On understanding grammar.* San Diego, CA: Academic Press.

Gleitman, L. R., Gleitman, H., & Shipley, E. F. (1972). The emergence of the child as grammarian. *Cognition, 1,* 137–164.

Goldhagen, D. J. (1996). *Hitler's willing executioners: Ordinary Germans and the Holocaust.* New York: Knopf.

Goodenough, W. H. (1956). Componential analysis and the study of meaning. *Language, 32,* 195–216.

Gottlieb, R. S. (1984). Mothering and the reproduction of power: Chodorow, Dinnerstein, and social theory. *Socialist Review, 14,* 93–119.

Gould, S. J. (1977). *Ontogeny and phylogeny.* Cambridge, MA: Harvard University Press.

Gould, S. J. (1980). Sociobiology and the theory of natural selection. In G. W. Barlow & J. Silverberg (Eds.), *Sociobiology: Beyond nature/nurture?* Washington, DC: American Association for the Advancement of Science.

Gould, S. J. (1981). *The mismeasure of man.* New York: W. W. Norton.

Gould, S. J. (1991). Not necessarily a wing. In *Bully for brontosaurus: Reflections in natural history* (pp. 139–151). New York: W. W. Norton.

Gould, S. J. (1993). *Eight little piggies.* New York: W. W. Norton.

Gould, S. J., & Vrba, E. S. (1982). Exaptation—a missing term in the science of form. *Paleobiology, 8,* 4–15.

Gregory, M. S., Silver, A., & Sutch, D. (Eds.). (1978). *Sociobiology and human nature.* San Francisco: Jossey-Bass.

Gregory, R. L. (1981). *Mind in science.* New York: Cambridge University Press.

Gumplowicz, L. (1885/1963). *Outlines of sociology.* New York: Paine-Whitman, 1963.

Hall, J. A. (1985). *Powers and liberties.* Oxford: Basil Blackwell.

Halpern, D. F. (1986). *Sex differences in cognitive abilities.* Hillsdale, NJ: Lawrence Erlbaum Associates.

Hamilton, W. D. (1964). The genetical evolution of social behavior. 1 & 2. *Journal of Theoretical Biology, 7,* 1–52.

Hebb, D. O. (1949). *The organization of behavior.* New York: Wiley.

Heider, K. G. (1976). Dani sexuality: A low energy system. *Man, 11,* 188–201.

Henderson, D. K. (1990). On the sociology of science and the continuing importance of epistemologically couched accounts. *Social Studies of Science, 20,* 113–148.

Hessen, B. M. (1931/1971). *The social and economic roots of Newton's 'Principia'.* New York: Fertig. (Originally published in *Science at the crossroads.* Kniga, London, 1931).

Hobson, J. A. (1988). *The dreaming brain.* New York: Basic Books.

Hobson, J. A., Hoffman, S. A., Helfand, R., & Kostner, D. (1987). Dream bizarreness and the activation-synthesis hypothesis. *Human Neurobiology, 6,* 157–164.

Hochberg, J. (1968). In the mind's eye. In R. N. Haber (Ed.), *Contemporary theory and research in visual perception.* New York: Holt.

Hughes, H. S. (1958). *Consciousness and society: The reorientation of European social thought 1890–1930.* New York: Knopf.

Hull, C. L., Hovland, C. I., Ross, R. T., Hall, M., Perkins, D. T., & Fitch, F. B. (1940). *Mathematico-deductive theory of rote learning; a study in scientific methodology.* New Haven, CT: Yale University Press.

Izard, C. E. (1972). *Patterns of emotion.* New York: Academic Press.

Izard, C. E. (1977). *Human emotions.* New York: Plenum Press.

Izard, C. E. (1992). Basic emotions, relations among emotions, and emotion-cognition relations. *Psychological Review, 99,* 561–565.

Izard, C. E., & Buechler, S. (1980). Aspects of consciousness and personality in terms of differential emotion theory. In R. Plutchik & H. Kellerman (Eds.), *Theories of Emotion.* New York: Academic Press.

Jackendoff, R. (1987). *Consciousness and the computational mind.* Cambridge, MA: MIT Press.

James, W. (1884). What is an emotion? *Mind, 9,* 188–205.

James, W. (1890). *The principles of psychology.* New York: Holt.

James, W. (1894). The physical basis of emotion. *Psychological Review, 1,* 516–529.

Jenkins, J. J. (1955). *Associative processes in verbal behavior: A report of the Minnesota conference.* Minneapolis: University of Minnesota, Department of Psychology.

Jensen, A. R. (1978). The current status of the IQ controversy. *Australian Psychologist, 13,* 7–28.

Joliot, M., Ribary, U., & Llinás, R. (1994). Human oscillatory brain activity near 40 Hz coexists with cognitive temporal binding. *Proceedings of the National Academy of Sciences of the U.S.A., 91,* 11748–11751.

Kahneman, D., & Treisman, A. (1984). Changing views of attention and automaticity. In R. Parasuraman & D. R. Davies (Eds.), *Varieties of attention* (pp. 29–61). New York: Academic Press.

Kamin, L. J. (1974). *The science and politics of IQ.* Potomac, MD: Lawrence Erlbaum Associates.

Kant, I. (1781/1929). *Critique of pure reason.* London: Macmillan.

Katona, G. (1940). *Organizing and memorizing.* New York: Columbia University Press.

Kierkegaard, S. A. (1844/1957). *The concept of dread.* Princeton, NJ: Princeton University Press.

Kingsolver, J. G., & Koehl, M. A. R. (1985). Aerodynamics, thermoregulation, and evolution of insect wings: Differential scaling and evolutionary change. *Evolution, 39,* 488–504.

Kitcher, P. (1982). *Abusing science: The case against creationism.* Cambridge, MA: MIT Press.

Kitcher, P. (1985). *Vaulting ambition.* Cambridge, MA: MIT Press.

Kitcher, P. (1990). Developmental decomposition and the future of human behavioral ecology. *Philosophy of Science, 57,* 96–117.

Kitcher, P. (1992). *Freud's dream: a complete interdisciplinary science of mind.* Cambridge, MA: MIT Press.

Kitcher, P. (1993). *The advancement of science.* New York: Oxford University Press.

Knecht, H., Pike-Tay, A., & White, R. (1993). *Before Lascaux.* Boca Raton, FL: CRC Press.

Köhler, W. (1917). *Intelligenzprüfungen an Anthropoiden.* Berlin: Königliche Akademie der Wissenschaften.

Köhler, W. (1925). *The mentality of apes.* New York: Harcourt, Brace.

Köhler, W. (1929). *Gestalt psychology.* New York: Liveright.

Köhler, W. (1938). *The place of value in a world of facts.* New York: Liveright.

Köhler, W. (1944). Value and fact. *The Journal of Philosophy, 41,* 197–212.

Kropotkin, P. A. (1902). *Mutual aid: A factor of evolution.* New York: McClure Phillips.

Lackner, J. R. (1988). Some proprioceptive influences on the perceptual representation of body shape and orientation. *Brain, 111,* 281–297.

Lashley, K. S. (1951). The problem of serial order in behavior. In L. A. Jeffress (Ed.), *Cerebral mechanisms in behavior: The Hixon symposium* (pp. 112–146). New York: Wiley.

Lavin, D. E., & Hyllegard, D. (1996). *Changing the odds.* New Haven, CT: Yale University Press.

Lazarus, R. S. (1991). *Emotion and adaptation.* New York: Oxford University Press.

Leacock, E. (1980). Social behavior, biology and the double standard. In G. W. Barlow & J. Silverberg (Eds.), *Sociobiology: Beyond nature/nurture?* Washington, DC: American Association for the Advancement of Science.

Lewontin, R. C. (1970). Race and intelligence. *Bulletin of the Atomic Scientists, 26,* 2–8.

Lieberman, P. (1989). The origins of some aspects of human language and cognition. In P. Mellars & C. Stringer (Eds.), *The human evolution.* Edinburgh: Edinburgh University Press.

Loftus, G. E., & Mackworth, N. H. (1978). Cognitive determinants of fixation location during picture viewing. *Journal of Experimental Psychology: Human Perception and Performance, 4,* 565–572.

Lorenz, K. (1963). *On aggression.* London: Metheun.

Lounsbury, F. G. (1956). A semantic analysis of the Pawnee kinship usage. *Language, 32,* 158–194.

Lumsden, C. J., & Wilson, E. O. (1981). *Genes, mind, and culture.* Cambridge, MA: Harvard University Press.

Lumsden, C. J., & Wilson, E. O. (1983). *Promethean fire.* Cambridge, MA: Harvard University Press.

Luria, S. E., Gould, S. J., & Singer, S. (1981). *A view of life.* Menlo Park, CA: Benjamin Cummings.

Lutz, C. (1983). Parental goals, ethnopsychology, and the acquisition of emotional meaning. *Ethos, 11,* 246–262.

Lutz, C. (1988). *Unnatural emotions: Everyday sentiments on a Micronesian atoll and their challenge to Western theory.* Chicago: University of Chicago Press.

Lutz, C., & Abu-Lughod, L. (1990). *Language and the politics of emotion.* Cambridge: Cambridge University Press.

Lycan, W. G. (1987). *Consciousness.* Cambridge, MA: MIT Press.

Maccoby, E. E., & Jacklin, C. N. (1974). *The psychology of sex differences.* Stanford, CA: Stanford University Press.

MacDowell, K. A., & Mandler, G. (1989). Constructions of emotion: Discrepancy, arousal, and mood. *Motivation and Emotion, 13,* 105–124.

MacIntyre, A. (1981). *After virtue: a study in moral theory.* London: Duckworth.

MacIntyre, A. (1988). *Whose justice? Which rationality?* London: Duckworth.

Mackenzie, B. (1980). Hypothesized genetic racial differences in IQ: A criticism of three proposed lines of evidence. *Behavior Genetics, 10,* 225–234.

Makkreel, R. A., & Rodi, F. (1989). Introduction. In R. A. Makkreel & F. Rodi (Eds.), *Wilhelm Dilthey—Selected works, Vol. I. Introduction to the human sciences* (pp. 3–43). Princeton, NJ: Princeton University Press.

Mandler, G. (1964). The interruption of behavior. In E. Levine (Ed.), *Nebraska symposium on motivation: 1964* (pp. 163–219). Lincoln: University of Nebraska Press.

Mandler, G. (1969). Acceptance of things past and present: A look at the mind and the brain. In R. B. McLeod (Ed.), *William James: Unfinished business*. Washington, DC: American Psychological Association.

Mandler, G. (1975a). Consciousness: Respectable, useful, and probably necessary. In R. Solso (Ed.), *Information processing and cognition: The Loyola symposium* (pp. 229–254). Hillsdale, NJ: Lawrence Erlbaum Associates.

Mandler, G. (1975b). *Mind and emotion*. New York: Wiley.

Mandler, G. (1979a). Emotion. In E. Hearst (Ed.), *The first century of experimental psychology* (pp. 275–321). Hillsdale, NJ: Lawrence Erlbaum Associates.

Mandler, G. (1979b). Organization, memory, and mental structures. In C. R. Puff (Ed.), *Memory organization and structure*. New York: Academic Press.

Mandler, G. (1982). The structure of value: Accounting for taste. In M. S. Clark & S. T. Fiske (Eds.), *Affect and cognition; The Seventeenth Annual Carnegie Symposium on Cognition*. Hillsdale, NJ: Lawrence Erlbaum Associates.

Mandler, G. (1984a). The construction and limitation of consciousness. In V. Sarris and A. Parducci (Eds.), *Perspectives in psychological experimentation: Toward the year 2000* (pp. 109–126). Hillside, NJ: Lawrence Erlbaum Associates.

Mandler, G. (1984b). *Mind and body: Psychology of emotion and stress*. New York: Norton.

Mandler, G. (1985). *Cognitive psychology: An essay in cognitive science*. Hillsdale, NJ: Lawrence Erlbaum Associates.

Mandler, G. (1988). Problems and directions in the study of consciousness. In M. Horowitz (Ed.), *Psychodynamics and cognition* (pp. 21–45). Chicago: Chicago University Press.

Mandler, G. (1989). Memory: Conscious and unconscious. In P. R. Solomon, G. R. Goethals, C. M. Kelley, & B. R. Stephens (Eds.), *Memory: Interdisciplinary approaches* (pp. 84–106). New York: Springer Verlag.

Mandler, G. (1990a). A constructivist theory of emotion. In N. S. Stein, B. L. Leventhal, & T. Trabasso (Eds.), *Psychological and biological approaches to emotion* (pp. 21–43). Hillsdale, NJ: Lawrence Erlbaum Associates.

Mandler, G. (1990b). Interruption (discrepancy) theory: Review and extensions. In S. Fisher & C. L. Cooper (Eds.), *On the move: The psychology of change and transition* (pp. 13–32). Chichester, England: Wiley.

Mandler, G. (1992a). Cognition and emotion: Extensions and clinical applications. In D. J. Stein & J. E. Young (Eds.), *Cognitive science and clinical disorders* (pp. 61–78). San Diego, CA: Academic Press.

Mandler, G. (1992b). Emotions, evolution, and aggression: Myths and conjectures. In K. T. Strongman (Ed.), *International Review of Studies on Emotion* (pp. 97–116). Chichester, England: John Wiley & Sons.

Mandler, G. (1992c). Toward a theory of consciousness. In H.-G. Geissler, S. W. Link, & J. T. Townsend (Eds.), *Cognition, information processing, and psychophysics: Basic issues* (pp. 43–65). Hillsdale, NJ: Lawrence Erlbaum Associates.

Mandler, G. (1993a). Approaches to a psychology of value. In M. Hechter, L. Nadel, & R. E. Michod (Eds.), *The origin of values* (pp. 229–258). Hawthorne, NY: Aldine de Gruyter.

Mandler, G. (1993b). Review of Dennett's "Consciousness explained." *Philosophical Psychology, 6,* 335–339.

Mandler, G. (1994a). Emotions and the psychology of freedom. In S. H. M. van Goozen, N. E. van de Poll, & J. A. Sergeant (Eds.), *Emotions: Essays on emotion theory* (pp. 241–262). Hillsdale, NJ: Lawrence Erlbaum Associates.

Mandler, G. (1994b). Hypermnesia, incubation, and mind-popping: On remembering without really trying. In C. Umiltà & M. Moscovitch (Eds.), *Attention and performance XV: Concious and nonconscious information processing* (pp. 3–33). Cambridge, MA: MIT Press.

Mandler, G. (1995). Origins and consequences of novelty. In S. M. Smith, T. B. Ward, & R. Finke (Eds.), *The creative cognition approach* (pp. 9–25). Cambridge, MA: MIT Press.

Mandler, G. (1996). Consciousness redux. In J. C. Cohen & J. W. Schooler (Eds.), *Scientific approaches to consciousness: The Twentyfifth Carnegie Symposium on Cognition* (pp. 479–498). Hillsdale, NJ: Lawrence Erlbaum Associates.

Mandler, G. (in press, a). Consciousness and mind as philosophical problems and psychological issues. In J. Hochberg and J. Cutting (Eds.), *Perception and cognition at century's end: History, philosophy, theory.* San Diego: Academic Press.

Mandler, G. (in press, b). Emotion. In D. E. Rumelhart & B. O. Martin (Eds.), *Cognitive Science.* San Diego, CA: Academic Press.

Mandler, G., & Kessen, W. (1974). The appearance of free will. In S. C. Brown (Ed.), *Philosophy of psychology* (pp. 305–324). London: Macmillan.

Mandler, G., & Shebo, B. J. (1982). Subitizing: An analysis of its component processes. *Journal of Experimental Psychology: General, 111,* 1–22.

Mandler, G., & Shebo, B. J. (1983). Knowing and liking. *Motivation and Emotion, 7,* 125–144.

Mandler, J. M. (1984a). Representation and recall in infancy. In M. Moscovitch (Ed.), *Infant memory* (pp. 75–101). New York: Plenum.

Mandler, J. M. (1984b). *Stories, scripts, and scenes: Aspects of schema theory.* Hillsdale, NJ: Lawrence Erlbaum Associates.

Mandler, J. M. (1988). How to build a baby: On the development of an accessible representational system. *Cognitive Development, 3,* 113–136.

Mandler, J. M. (1992). How to build a baby II: Conceptual primitives. *Psychological Review, 99,* 587–604.

Mandler, J. M., & Johnson, N. S. (1976). Some of the thousand words a picture is worth. *Journal of Experimental Psychology: Human Learning and Memory, 2,* 529–540.

Mandler, J. M., & Mandler, G. (1964). *Thinking: From association to Gestalt.* New York: Wiley.

Mandler, J. M., & Mandler, G. (1968). The diaspora of experimental psychology: The Gestaltists and others. In D. Fleming & B. Bailyn (Eds.), *The intellectual migration: Europe and America, 1930–1960.* Cambridge, MA: Charles Warren Center, Harvard University.

Mandler, P. (1989, Winter). The "double life" in academia. *Dissent,* 94–99.

Mann, M. (1986). *The sources of social power.* Cambridge: Cambridge University Press.

Marcel, A. J. (1983a). Conscious and unconscious perception: An approach to the relations between phenomenal experience and perceptual processes. *Cognitive Psychology, 15,* 238–300.

Marcel, A. J. (1983b). Conscious and unconscious perception: Experiments on visual masking and word recognition. *Cognitive Psychology, 15,* 197–237.

Marden, J. H., & Kramer, M. G. (1994). Surface-skimming stoneflies: A possible intermediate stage in insect flight evolution. *Science, 266,* 427–430.

Marlowe, J. (1971). *The golden age of Alexandria*. London: Victor Gollancz.

Mayr, E. (1982). *The growth of biological thought: Diversity, evolution, and inheritance*. Cambridge, MA: Harvard University Press.

McClelland, J. L., & Rumelhart, D. E. (1981). An interactive activation model of context effects in letter perception: Part 1. An account of basic findings. *Psychological Review, 88*, 375–407.

McGeoch, J. A. (1942). *The psychology of human learning, an introduction*. New York: Longmans, Green.

McGue, M., & Lykken, D. T. (1992). Genetic influence on risk of divorce. *Psychological Science, 3*, 368–373.

McNaughton, M. (1989). *Biology and emotion*. Cambridge: Cambridge University Press.

Mechanisation of thought processes. (1959). (Symposium No. 10, National Physical Laboratory). London: Her Majesty's Stationery Office.

Meyer, L. B. (1956). *Emotion and meaning in music*. Chicago: University of Chicago Press.

Miller, G. A. (1956). The magical number seven, plus or minus two: Some limits on our capacity for processing information. *Psychological Review, 63*, 81–97.

Miller, G. A. (1985). The constitutive problem of psychology. In S. Koch & D. E. Leary (Eds.), *A century of psychology as science*. New York: McGraw Hill.

Miller, G. A., Galanter, E. H., & Pribram, K. (1960). *Plans and the structure of behavior*. New York: Holt.

Mivart, S. G. J. (1871). *On the genesis of species*. New York: D. Appleton.

Molony, C. H., & Tuan, D. (1976). Further studies on the Tasaday language: Texts and vocabulary. In D. E. Yen & J. Nance (Eds.), *Further studies on the Tasaday*. Makati, Philippines: Panamin Foundation.

Muenzinger, K. F. (1938). Vicarious trial and error at a point of choice. I. A general survey of its relation to learning efficiency. *Journal of Genetic Psychology, 53*, 75–86.

Nagel, T. (1988). Agreeing in principle. *The Times Literary Supplement, No. 4449*, 747–748.

Nance, J. (1975). *The gentle Tasaday*. New York: Harcourt Brace Jovanovich.

Neisser, U. (1967). *Cognitive psychology*. New York: Appleton-Century-Crofts.

Neisser, U., Boodoo, G., Bouchard, T. J., Jr., Boykin, A. W., Brody, N., Ceci, S. J., Halpern, D. F., Loehlin, J. C., Perloff, R., Sternberg, R. J., & Urbina, S. (1996). Intelligence: Knowns and unknowns. *American Psychologist, 51*, 77–101.

Newell, A., & Simon, H. A. (1956). The logic theory machine: A complex information processing system. *IRE Transactions on information theory, IT-2(3)*, 61–79.

Nielsen, T. I. (1963). Volition: A new experimental approach. *Scandinavian Journal of Psychology, 4*, 225–230.

Nisbett, R. E., & Wilson, T. D. (1977). Telling more than we can know: Verbal reports on mental processes. *Psychological Review, 84*, 231–259.

Oatley, K. (1992). *Best laid schemes: The psychology of emotions*. New York: Cambridge University Press.

Oatley, K., & Johnson-Laird, P. N. (1987). Towards a cognitive theory of emotion. *Cognition & Emotion, 1*, 29–50.

Ogbu, J. U. (1978). *Minority education and caste: The American system in cross-cultural perspective*. New York: Harcourt Brace Jovanovich.

Ogbu, J. U. (1994). From cultural differences to differences in cultural frame of reference. In P. M. Greenfield & R. R. Cocking (Eds.), *Cross-cultural roots of minority child development* (pp. 365–391). Hillsdale, NJ.: Lawrence Erlbaum Associates.

Oppenheim, F. E. (1968). Freedom. In D. L. Sills (Ed.), *International encyclopedia of the social science* (pp. 554–559). New York: Macmillan.

Ortony, A. (1991). Value and emotion. In W. Kessen, A. Ortony, & F. Craik (Eds.),

Memories, thoughts, emotions: Essays in honor of George Mandler (pp. 337–353). Hillsdale, NJ: Lawrence Erlbaum Associates.

Ortony, A., Clore, G. L., & Collins, A. (1988). *The cognitive structure of emotions.* New York: Cambridge University Press.

Ortony, A., & Turner, T. J. (1990). What's basic about basic emotions? *Psychological Review, 97,* 315–331.

Pagels, E. H. (1988). *Adam, Eve, and the serpent.* New York: Random House.

Panksepp, J. (1992). A critical role for "affective neuroscience" in resolving what is basic about basic emotions. *Psychological Review, 99,* 554–560.

Partridge, P. H. (1967). Freedom. In P. Edwards (Ed.), *The encyclopedia of philosophy* (pp. 221–225). New York: Macmillan.

Paulhan, F. (1887). *Les phénomènes affectifs et les lois de leur apparition.* Paris: F. Alcan.

Perry, R. B. (1926). *General theory of value: Its meaning and basic principles construed in terms of interest.* Cambridge, MA: Harvard University Press.

Piaget, J. (1953). *The origin of intelligence in the child.* London: Routledge.

Piaget, J. (1954). *The construction of reality in the child.* New York: Basic Books.

Pick, J. (1970). *The autonomic nervous system.* Philadelphia: Lippincott.

Pickering, A. (1984). *Constructing quarks: A sociological history of particle physics.* Chicago: University of Chicago Press.

Pinker, S. (1994). *The language instinct.* New York: W. Morrow.

Pinker, S., & Bloom, P. (1990). Natural language and natural selection. *Behavioral and Brain Sciences, 13,* 707–784.

Pippard, B. (1985). Discontinuities. *London Review of Books, 7,* 8–9.

Plomin, R. (1990). *Nature and nurture: An introduction to human behavioral genetics.* Pacific Grove, PA: Brooks Cole.

Plomin, R., Corley, R., DeFries, J. C., & Fulker, D. W. (1990). Individual differences in television viewing in early childhood. *Psychological Science, 1,* 371–377.

Plutchik, R. (1980). *Emotion: A psychoevolutionary synthesis.* New York: Harper & Row.

Pollack, R. H., & Brenner, M. W. (1969). *The experimental psychology of Alfred Binet: Selected papers.* New York: Springer.

Popkin, S. L. (1991). *The reasoning voter: Communication and persuasion in presidential campaigns.* Chicago: University of Chicago Press.

Porter, T. M. (1986). *The rise of statistical thinking 1820–1900.* Princeton, NJ: Princeton University Press.

Posner, M. I., & Snyder, C. R. R. (1975). Attention and cognitive control. In R. Solso (Ed.), *Information processing and cognition: The Loyola symposium.* Potomac, MD: Lawrence Erlbaum Associates.

Postman, N. (1988). Social science as moral theology. In *Conscientious objections.* New York: Alfred A. Knopf.

Prescott, C. A., Johnson, R. C., & McArdle, J. J. (1991). Genetic contributions to television viewing. *Psychological Science, 2,* 430–431.

Purcell, T. (1986). [Typicality, interestingness, and preference]. University of Sydney. Unpublished data.

Putnam, H. (1980). Philosophy and our mental life. In N. Block (Ed.), *Readings in the philosophy of psychology* (Vol. 1,). Cambridge, MA: Harvard University Press.

Reber, A. (1996). How to differentiate implicit and explicit modes of acquisition. In J. D. Cohen & J. W. Schooler (Eds.), *Scientific approaches to consciousness: The Twenty-fifth Annual Carnegie Symposium on Cognition.* Hillsdale, NJ: Lawrence Erlbaum Associates.

Robert, W. (1886). *Der Traum als Naturnothwendigkeit erklärt.* Hamburg: H. Seippel.

Rokeach, M. (1973). *The nature of human values.* New York: The Free Press.

Rosch, E. (1978). Principles of categorization. In E. Rosch & B. B. Lloyd (Eds.), *Cognition and categorization*. Hillsdale, NJ: Lawrence Erlbaum Associates.

Rosen, C. (1971). *The classical style*. New York: Viking Press.

Rumelhart, D. E., & McClelland, J. L. (1985). *Parallel distributed processing: Explorations on the microstructure of cognition*. Cambridge, MA: MIT Press.

Rumelhart, D. E., & Ortony, A. (1978). The representation of knowledge in memory. In R. C. Anderson, R. J. Spiro, & W. E. Montague (Eds.), *Schooling and the acquisition of knowledge*. Hillsdale, NJ: Lawrence Erlbaum Associates.

Ruskin, J. (1843/1906). *Modern Painters*, vol. 3. London: J. M. Dent Co.

Russell, J. A. (1994). Is there universal recognition of emotion from facial expressions? A review of the cross-cultural studies. *Psychological Bulletin, 115*, 102–141.

Russell, J. A. & Fernández-Dols, J. M. (1997). *The psychology of facial expression*. New York: Cambridge University Press.

Schachter, S. (1970). The assumption of identity and peripheralist-centralist controversies in motivation and emotion. In M. B. Arnold (Ed.), *Feelings and emotions*. New York: Academic Press.

Schachter, S., & Singer, J. E. (1962). Cognitive, social and physiological determinants of emotional state. *Psychological Review, 69*, 379–399.

Schacter, D. L. (1987). Implicit memory: History and current status. *Journal of Experimental Psychology: Learning, Memory, and Cognition, 13*, 501–518.

Schnabel, F. (1950). *Deutsche Geschichte im Neunzehnten Jahrhundert* (2nd ed.). Freiburg im Breisgau, Germany: Verlag Herder.

Schwartz, B. (1986). *The battle for human nature*. New York: Norton.

Schweder, R. A. (1982). Beyond self-constructed knowledge: The study of culture and morality. *Merrill-Palmer Quarterly, 28*, 41–69.

Searle, J. R. (1983). *Intentionality*. New York: Cambridge University Press.

Searle, J. R. (1992). *The rediscovery of the mind*. Cambridge, MA: MIT Press.

Selz, O. (1913). *Über die Gesetze des geordneten Denkverlaufs. Eine experimentelle Untersuchung*. Stuttgart: Spemann.

Shallice, T. (1988). *From neuropsychology to mental structure*. Cambridge: Cambridge University Press.

Shallice, T. (1991). The revival of consciousness in cognitive science. In W. Kessen, A. Ortony, & F. Craik (Eds.), *Memories, thoughts, emotions: Essays in honor of George Mandler* (pp. 213–226). Hillsdale, NJ: Lawrence Erlbaum Associates.

Shanks, D. R., Green, R. E., & Kolodny, J. A. (1994). A critical examination of the evidence for unconscious (implicit) learning. In C. Umiltà & M. Moscovitch (Eds.), *Attention and Performance XV: Conscious and nonconscious information processing* (pp. 837–860). Cambridge, MA: MIT Press.

Sheehan, J. J. (1978). *German liberalism in the nineteenth century*. Chicago: University of Chicago Press.

Silverberg, J. (1978). The scientific discovery of logic: The anthropological significance of empirical research on psychic unity (inference-making). In M. D. Loflin & J. Silverberg (Eds.), *Discourse and inference in cognitive anthropology: An approach to psychic unity and enculturation*. The Hague: Mouton.

Silverberg, J. (1980). Sociobiology, the new synthesis? An anthropologist's perspective. In G. W. Barlow & J. Silverberg (Eds.), *Sociobiology: Beyond nature/nurture?* Washington, DC: American Association for the Advancement of Science.

Slater, M. K. (1959). Ecological factors in the origin of incest. *American Anthropologist, 61*, 1042–1059.

Snygg, D., & Combs, A. W. (1949). *Individual behavior*. New York: Harper.

Sowell, T. (1986). *A conflict of visions*. New York: William Morrow.

Staub, E. (1996). Cultural-societal roots of violence. *American Psychologist, 51*, 117–132.

Stern, F. (1961). *The politics of cultural despair: A study in the rise of the Germanic ideology.* Berkeley, CA: University of California Press.

Sternberg, R. J. (1985). *Beyond IQ: a triarchic theory of human intelligence.* New York: Cambridge University Press.

Stringer, C., & Gamble, C. (1993). *In search of the Neanderthals: Solving the puzzle of human origins.* New York: Thames & Hudson.

Strum, S. C. (1987). *Almost human.* New York: Random House.

Taylor, F. W. (1911). *The principles of scientific management.* New York: Harper & Brothers.

Thagard, P. (1986). Parallel computation and the mind-body problem. *Cognitive Science, 10*, 301–318.

Thwaites, R. G. (1906). *The Jesuit relations and allied documents.* Cleveland, OH: Burrows.

Tobias, P. V. (1990). Some critical steps in the evolution of the hominid brain. *Pontificae Academiae Scientiarum Scripta Varia, 78*, 741–761.

Tobias, P. V. (1995). *The communication of the dead: Earliest vestiges of the origin of articulate language.* Seventeenth Kroon Lecture, Stichting Nederlands Museum voor Anthropologie en Praehistorie, Amsterdam.

Toda, M. (1982). *Man, robot, and society.* Boston: Martinus Nijhoff.

Tomkins, S. (1962–1992). *Affect, imagery, and consciousness.* New York: Springer.

Tomkins, S. S. (1981). The quest for primary motives: Biography and autobiography of an idea. *Journal of Personality and Social Psychology, 41*, 306–329.

Treisman, A. M. (1964). Verbal cues, language and meaning in selective attention. *American Journal of Psychology, 77*, 206–218.

Tuchman, B. W. (1966). *The proud tower: A portrait of the world before the war 1890–1914.* New York: Macmillan.

Tulving, E. (1985). Memory and consciousness. *Canadian Psychology, 26*, 1–12.

Turing, A. M. (1950). Computing machinery and intelligence. *Mind, 49*, 433–460.

Turner, T. J., & Ortony, A. (1992). Basic emotions: Can conflicting criteria converge? *Psychological Review, 99*, 566–571.

Tversky, A., & Kahneman, D. (1973). Availability: A heuristic for judging frequency and probability. *Cognitive Psychology, 5*, 207–232.

Van Orden, G. & Uyeda, K. (1984). [Typicality and preference]. Arizona State University. Unpublished data.

von Hartmann, E. (1869). *Philosophie des Unbewussten.* Berlin: Duncker.

Wald, G. (1978). The human condition. In M. S. Gregory, A. Silver, & D. Sutch (Eds.), *Sociobiology and human nature* (pp. 277–282). San Francisco: Jossey-Bass.

Washburn, S. L. (1978). Animal behavior and social anthropology. In M. S. Gregory, A. Silver, & D. Sutch (Eds.), *Sociobiology and human nature* (pp. 53–74). San Francisco: Jossey-Bass.

Watson, J. B. (1913). Psychology as the behaviorist views it. *Psychological Review, 20*, 158–177.

Watson, J. B. (1919). *Psychology from the stand-point of a behaviorist.* Philadelphia: Lippincott.

Watt, H. J. (1905–1906). Experimental contribution to a theory of thinking. *Journal of Anatomy and Physiology, 40*, 257–266.

Webster's Seventh New Collegiate Dictionary. (1969). Springfields, MA: G & C Merriam Company.

Weiskrantz, L. (1985). On issues and theories of the human amnesic syndrome. In N. M. Weinberger, J. L. McGaugh, & G. Lynch (Eds.), *Memory systems of the brain: Animal and human cognitive processes* (pp. 380–415). New York: Guilford Press.

Wiebe, R. H. (1967). *The search for order: 1877–1920.* New York: Hill and Wang.

Wilson, D. S., & Sober, E. (1994). Reintroducing group selection to the human behavioral sciences. *Behavioral and Brain Sciences, 17,* 585–654.

Wundt, W. (1874). *Grundzüge der physiologischen Psychologie.* Leipzig: W. Engelmann.

Wundt, W. (1896). *Grundriss der Psychologie.* Leipzig: Wilhelm Engelmann.

Wundt, W. (1900–1909). *Völkerpsychologie: Eine Untersuchung der Entwicklungsgesetze von Sprache, Mythus und Sitte.* Leipzig: W. Engelmann.

Wundt, W. (1908–1911). *Grundzüge der physiologischen Psychologie* (6th ed.). Leipzig: W. Engelmann.

Wundt, W. (1912). *Elemente der Völkerpsychologie: Grundlinien einer psychologischen Entwicklungsgeschichte der Menschheit.* Leipzig: A. Kröner Verlag.

Wundt, W. (1916). *Elements of folk psychology: Outline of a psychological history of the development of mankind.* London: Allen & Unwin.

Wundt, W. (1920). *Erlebtes und Erkanntes.* Stuttgart: Alfred Kröner Verlag.

Yen, D. E. (1976). The ethnobotany of the Tasaday: III. Notes on the subsistence system. In D. E. Yen & J. Nance (Eds.), *Further studies on the Tasaday.* Makati, Philippines: Panamin Foundation.

Yen, D. E., & Nance, J. (Eds.). (1976). *Further studies on the Tasaday.* Makati, Philippines: Panamin Foundation.

Zajonc, R. B. (1968). Attitudinal effects of mere exposure. *Journal of Personality and Social Psychology Monograph, 9,* 1–28.

Zajonc, R. B. (1980). Feeling and thinking: Preferences need no inferences. *American Psychologist, 35,* 151–175.

Zajonc, R. B. (1984). On the primacy of affect. *American Psychologist, 39,* 117–123.

Zborowski, M. (1969). *People in pain.* San Francisco: Jossey-Bass.

Index of Names

Abu-Lughod, L., 76–77
Ahmed, S. Z., 121
Alexander the Great, 12
Anaximander, 11
Andrews, P., 140
Aquinas, Saint Thomas, 13
Aristarchus of Samos, 12
Aristotle, 11, 13, 68–69
Arnheim, R., 93–94, 96
Augustine, Saint, 12, 124–125
Averill, J. R., 82
Avicenna, 13

Baars, B. J., 164
Baillargeon, R., 18
Bain, A., 156, 159
Baldwin, J. M., 157
Bandura, A., 108
Barash, D. P., 25, 33, 119–120
Bareh, H., 120–121
Barlow, G. W., 32–33
Bartlett, F. C., 163
Bates, E., 140

Bavelas, J. B., 81
Beach, F. A., 33, 118
Bebel, A., 150
Bem, D. J., 84, 90
Berger, J., 96
Berlin, I., 128–130, 133
Berscheid, E., 36–37, 73, 76, 132
Binet, A., 113, 159, 163
Black, A., 81
Bloom, P., 140
Blumenthal, A. L., 152, 154
Boodoo, G., 113
Bouchard, T. J., Jr., 113, 114–115
Boykin, A. W., 113
Bracha, H. S., 38
Brenner, M. W., 163
Brickman, P., 91
Briggs, J. L., 108
Broadbent, D. E., 163–164
Brody, N., 113
Bruce, D., 163
Bruner, J., 164
Buechler, S., 79

Burkert, W., 3
Burks, A., 47
Buss, A. R., 148
Buss, D. M., 121

Cannon, W. B., 69, 70, 93
Carey, S., 17
Ceci, S. J., 113
Chagnon, N. A., 111
Chapman, A. J., 6
Chomsky, N., 49, 128, 140–141, 164
Claparède, E., 163
Clore, G. L., 67, 72, 79–80
Cofer, C. N., 163–164
Cole, M., 52
Coleridge, H., 129
Collins, A., 67, 72, 79–80
Combs, A. W., 163
Copernicus, N., 13
Corley, R., 37
Craik, K. J. W., 163
Crandall, R., 91
Crick, F., 59–60

D'Andrade, R., 57
Darwin, C., 14, 16, 22–23, 27, 80, 150,
  156–157
Davis, J. O., 38
Dawkins, R., 24
Deese, J., 166
DeFries, J. C., 37
deGroot, A. D., 159
Dellas, M., 95
Democritus, 11
Dennett, D. C., 43, 64, 165
Dentan, R. K., 108–109
Descartes, R., 13–14, 43, 69
Deutsch, K. W., 41
deWaal, F. B. M., 24, 86, 106, 125
Dewey, J., 49, 69
Diamond, R., 17
Dilthey, W., 151
Dollard, J., 104
Donald, M., 138
Doob, L. W., 104
Du Bois-Reymond, E., 150, 152
Duncker, K. 159
Durkheim, E., 155
Dworkin, R., 130

Edison, T., 158
Eibl-Eibesfeldt, I., 110
Ekman, P., 81–82
Eldredge, N., 28, 31, 144
Elizalde, M. J., 109–110
Epictetus, 129
Eratosthenes, 12
Esper, E. A., 161
Euclid, 12

Fechner, G. T., 151, 154
Fernandez, C. A., II, 109–110
Fernández-Dols, J. M., 80
Fischman, J., 28
Fitch, F. B., 163
Fleck, L., 148
Flynn, J. R., 115–116
Fodor, J., 49, 165
Folkman, S., 80
Ford, C. S., 118
Fox, R. B., 109
Freud, S., 4, 15, 41, 59, 93, 104, 134, 158,
  167
Fridlund, A. J., 67, 80–81
Friedman, A., 70
Frijda, N. H., 79, 159
Fromm, E., 103
Fulker, D. W., 37

Gaier, E. L., 95
Galanter, E. H., 164
Galileo Galilei, 13
Gamble, C., 30
Gardner, H., 113
Garner, W., 162
Gaver, W., 91
Gay, J., 52
Geertz, C., 67–68, 86
Gellner, E., 135
Gentner, D., 141
Gergen, K. J., 165
Gibson, J. J., 89
Givòn, T., 140–141
Gleitman, H., 63
Gleitman, L. R., 63
Glick, J., 52
Godwin, W., 5
Goldhagen, D. J., 126
Goodenough, W. H., 164

Gottlieb, R. S., 100–101
Gould, S. J., 5, 21, 26, 32, 34, 62, 114, 141
Green, R. E., 51
Gregory, M. S., 32
Gregory, R. L., 54
Gumplowicz, L., 101
Gutenberg, J., 13

Hall, J. A., 128, 133–134
Hall, M., 163
Halpern, D. F., 113, 119–120
Hamilton, W. D., 23
Harrison, A. A., 91
Hayek, F.-A., 168
Hebb, D. O., 162
Heider, K. G., 119
Helfand, R., 59
Helmholtz, H., 150, 152
Henderson, D. K., 148
Herophilus, 12
Herwegh, G., 149
Hessen, B. M., 148
Hirsh, I., 182
Hobson, J. A., 59–60
Hochberg, J., 164
Hoffman, S. A., 59
Hovland, C. I., 163
Hughes, H. S., 158
Hull, C. L., 161, 163
Hyllegard, D., 118

Izard, C. E., 79, 81–82

Jackendoff, R., 43, 47, 54, 165
Jacklin, C. N., 119
James, W., 67–68, 81, 126, 156–157, 160
Jenkins, J. J., 164
Jensen, A. R., 113
Johnson, N. S., 70
Johnson, R. C., 37
Johnson-Laird, P. N., 80
Joliot, M., 59
Jones, D. M., 6
Jones, S., 38
Jovanian, 124

Kahneman, D., 5, 57, 168
Kamin, L. J., 36
Kant, I., 19, 43–44, 163

Katona, G., 137, 159, 162–163
Kessen, W., 130
Kierkegaard, S. A., 130
King, M. L., 132
Kingsolver, J. G., 27
Kitcher, Patricia, 153, 167
Kitcher, Philip, 21, 32, 34, 83, 148
Knecht, H., 29
Koehl, M. A. R., 27
Köhler, W., 85, 89, 159, 163
Koffka, K., 159
Kolodny, J. A., 51
Kostner, D., 59
Kramer, M. G., 27
Kropotkin, P. A., 24

Lackner, J. R., 57
Lamarck, J.B., 23
Lashley, K. S., 163
Lassalle, A., 150
Lavin, D. E., 118
Lazarus, R. S., 80
Leacock, E., 111, 119–120
Leibniz, G. W., 14
Lemery, C. R., 81
Lewin, K., 159
Lewontin, R. C., 113
Licklider, J.C.R., 162
Lieberman, P., 29
Llinàs, R., 59
Locke, J., 135
Loeb, J., 161
Loehlin, J. C., 113
Loftus, G. E., 70
Lorenz, K., 32, 104
Lounsbury, F. G., 164
Lumsden, C. J., 15–16, 19, 110–111
Luria, S. E., 21
Luther, M. 13
Lutz, C., 76–78
Lycan, W. G., 41
Lykken, D. T., 37, 114–115
Lynch, F. S. J., 109–110

Maccoby, E. E., 119
MacDowell, K. A., 71
MacIntyre, A., 126–127
Mackenzie, B., 113
Mackworth, N. H., 70

Makkreel, R. A., 151
Mandler, G., *passim*
Mandler, J. M., 16, 44, 50, 63–64, 70, 93,
    139, 142, 158, 159, 161
Mandler, P., 99
Mann, M., 128, 133
Marcel, A. J., 47
Marchman, V., 140
Marden, J. H., 27
Marlowe, J., 12
Marx, K., 6, 15, 98, 134, 135, 150
Mayr, E., 29
McArdle, J. J., 37
McClelland, J. L., 53–54
McGeoch, J. A., 49, 157
McGue, M. 37, 114–115
McNaughton, M., 67
Mead, G. H., 167
Mendel, G., 23
Meyer, L. B., 96–97
Mill, J. S., 128, 151
Miller, N. E., 104
Miller, G. A., 17, 46, 52, 162, 164
Mitchison, G., 59–60
Mivart, S. G. J., 27
Molony, C. H., 109
Morgan, C., 162
Mowrer, O. H., 104
Muenzinger, K. F., 61
Müller, J., 150, 152
Mullett, J., 81

Nagel, T., 127
Nance, J., 109–110
Neisser, U., 113, 164
Newell, A., 164
Newton, I., 14
Nielsen, T. I., 57
Nisbett, R. E., 57

Oatley, K., 69, 80
Ogbu, J. U., 116–118
Oppenheim, F. E., 129
Ortony, A., 44, 67, 72, 79–80, 82, 89, 163

Pagels, E. H., 13, 125
Panksepp, J., 82
Parsons, T., 167
Partridge, P. H., 128–129, 133
Pascal, B., 14

Paulhan, F., 69
Perkins, D. T., 163
Perloff, R., 113
Perry, R. B., 86
Phelps, J. A., 38
Piaget, J., 18, 45, 157, 163–164
Pick, J., 83
Pickering, A., 148
Pike-Tay, A., 29
Pinker, S., 140–141, 144–145
Pippard, B., 64, 143
Plato, 11
Plomin, R., 36–37
Plutchik, R., 82
Pollack, R. H., 163
Popkin, S. L., 168
Porter, T. M., 152, 154
Posner, M. I., 54
Postman, N., 166
Prescott, C. A., 37
Pribram, K., 164
Ptolemy, 13
Purcell, T., 93
Putnam, H., 64, 143–144
Pythagoras, 12

Reber, A., 51
Redfield, J., 91
Ribary, U., 59
Robert, W., 60
Rodi, F., 151
Rokeach, M., 88–89
Rosch, E., 93
Rosen, C., 96–97
Rosenblith, W., 162
Ross, R. T., 163
Rousseau, J. J., 129, 135
Rumelhart, D. E., 44, 53–54, 163
Ruskin, J., 67
Russell, J. A., 80

Schachter, S., 69, 164
Schacter, D. L., 51
Schnabel, F., 152
Schwartz, B., 161
Schweder, R. A., 86
Searle, J. R., 41, 47
Sears, R. R., 104
Selz, O., 158–159
Shallice, T., 46, 69

Shanks, D. R., 51
Shannon, C. E., 162
Shebo, B. J., 52, 79, 91–92
Sheehan, J. J., 150, 153
Shipley, E. F., 63
Siegal, N. L., 114–115
Silver, A., 32
Silverberg, J., 15, 18, 32–33
Simon, H. A., 164
Singer, J. E., 69, 164
Singer, S., 21
Skinner, B. F., 158, 161
Slater, M. K., 15
Smith, A., 5, 124
Snyder, C. R. R., 54
Snygg, D., 162
Sober, E., 25
Socrates, 11
Sontag, S., 53
Sowell, T., 5–6
Spelke, E. S., 18
Spencer, H., 157
Spinoza, B., 129
Staub, E., 108
Stern, F., 152
Sternberg, R. J., 113
Stevens, A., 141
Stevens, S. S., 162
Stringer, C., 30, 140
Strum, S. C., 86, 105–106
Sutch, D., 32

Tattersall, I., 28, 31, 144
Taylor, F. W., 161
Tellegen, A., 114–115
Thagard, P., 47, 155
Thal, D., 140
Thales, 11
Thwaites, R. G., 111
Titchener, E. B., 156–157
Tobias, P. V., 28–29, 141
Toda, M., 82
Tolman, E. C., 161

Tomkins, S. S., 81
Treisman, A., 57, 164
Tuan, D., 109
Tuchman, B. W., 153, 157, 160
Tulving, E., 61
Turing, A. M., 14, 43, 161
Turner, T. J., 67, 82
Tversky, A., 5, 168

Urbina, S., 113
Uyeda, K., 93

van Gulick, R., 47
Van Orden, G., 93
Vives, J. L., 13
Voltaire, 25
von Hartmann, E., 159
von Neumann, J., 161–162
Vrba, E. S., 26

Wald, G., 82
Washburn, S. L., 32–33
Wasserman, S., 18
Watson, J. B., 160
Watt, H. J., 159
Weaver, W., 162
Weber, E. H., 151
Weiskrantz, L., 60
Wertheimer, M., 159
White, R., 29
Whitehead, A. N., 64, 143
Wiebe, R. H., 160
Wiener, N., 162
Wilson, D. S., 25
Wilson, E. O., 15–16, 19, 110–111
Wilson, T. D., 57
Wundt, W., 140, 149–156, 166

Yen, D. E., 109–110

Zajonc, R. B., 78–79, 91
Zborowski, M., 42

Index of Subjects

Activation and consciousness, 54–56
Adam and Eve, sin of, 12, 124
  as models of submissiveness, 124–125
  sociobiology and, 12
Adaptation, 24–26
  Panglossian view, 25–26
Affect, *see also* emotion
  and emotion contrasted, 68
Aggression, 101–111, *see also* cooperation
  as adaptive, 104
  and childrearing, 108–110
  defensive, 103
  definition of, 102
  derived from thwarting, 107–108
  as drive, 104
  parental, 103
  as primate characteristic, 105
    variations of, 105–106
Arts and values, 96–97
Assimilation and accommodation, 45, 92, 94
Associations, labeled, 139
Astronomy, as model for psychology, 166

Behaviorism, 160–161
  and American puritanism, 161

Church history, 12–13
  early disputes on sexuality, 124
Cognitive psychology, sources of, 162–164
Cognitive science, 165
Conscious contents, in daily life, 56–59
  and current knowledge, 58
  in dreams, 59–60
  and emotions, 58
  and veridicality, 57
Consciousness, 46–65, *see also* Limited capacity, conscious contents
  altered states of, 58–59
  and amnesia, 59
  of animals, 63
  and choice, 51
  and cognitive complexity, 62–63
  without constraints, 53
  constructed, 48
  as epiphenomenon, 47

Consciousness (*continued*)
    evolution of, 61–65
    as executive, 47
    feedback function of, 54–55
    and intentionality, 48
    and learning, 50, 53
    and memory, 60–61
    neurophysiology of, 64–65
    and self, 62
    serial nature of, 52–53
Constraints, on human nature, 7, 9–20
    conditions of acceptance of, 133
    biological, 7, 15–20
    categories of social, 131–132
    emotional consequences of, 130
    and freedom, 123, *see also* freedom
    historical, 9–15
    social, 7–8
Contradictions, 92–93, *see also*
        discrepancies
Cooperation, as alternative to aggression,
        108–110
    practiced by hunter-gatherers, 110–111

Determinist vs. Humanist views of
        psychology, 168
Difference detection, 70–71
    function of, 70
    and values, 91–93
Discrepancies, *see also* difference detection
    and the arts, 96–97
    and freedom, 130
    and novelty, 95–96
    ubiquity of, 74–75
    and values, 91–92
Dreams, *see* conscious contents

Emotion, 66–84, *see also* value
    adaptive value of, 74
    basic, evidence for, 81–82
    categories of, 67, 72, 78
    construction of, 71–74
    definitions of, 67–69
    evolution of, 82–84
    of freedom, sources of, 132
    grief and guilt compared, 78
    hostility, sources of, 107
    and joy of completion, 94
    and liberty, 132
    overdetermination of, 74

positive and negative, 75–76
    of simple stimuli, 56
    sources of contents, 51–52
Environments, measuring of, 38
    and child rearing, 121–122
Evaluation,
    cultural determinants of, 76–78
    sources of, 72–73
Evolution, 21–32
    and alternate theories of selection, 23
    and change, 21–22, 31–32
    climate and gravity, 20
    of complex behavior, 34
    of cooperation, 39
    and geography, 22
    of humans, 28–30
    and intelligence, 112
    of language, 140–141
    and natural selection, 22–23, 24–26
    parallel, 22, 31, 141
    and speech and language, 28–29
Exaptation, 26–28
    and evolution of emotion, 83–84
    and evolution of wings, 27–28
    and language, 141
    and preadaptation, 26

Face recognition, 17
Facial expressions, 16
    and emotion, 80–81
Familiarity, 91
Fin de siècle, culture of, 157–158
    and *Gestalt* psychology, 159
Freedom, 127–133
    definitions of, 129
    and constraints, 128–133
    natural, 128
    natural and constructed, contrasted,
        129
Functionalism, definitions of, 49
    in psychology, 157, 160

Geisteswissenschaft and *Naturwissenschaft*,
        151, 153, 155–156, 166
    contrasted, 151
Gender differences, *see* sex differences
Genetic mechanisms,
    and adoptive twin studies, 34–37,
        114–115
    and environmental interactions, 36–37

Mendelian, 23
overgeneralization of, 145
proximate vs. distal analyses of, 34, 37, 83, 98, 107–108
in schizophrenia, 38

Homeostasis, 70
Homology and analogy, 31
in culture, 77
Human nature,
cultural views of, 4
seen as culture free, 4
dimensions of, 6–7
divided religious views of, 124
and mental mechanisms, 137–138
positive views of, 5

Information age, and psychology, 161–164
Intelligence tests, 113
increase over time, 115–116
limitations of, 113
socio-economic bias of, 117
Intelligence, 112–118
and caste-like environments, 116–118
multidimensionality of, 114
social basis of, 115–116
and twin studies, 114–115

James, and functionalism, 156–157
and Wundt contrasted, 156

Khasi matrilineal culture, 120–121

Language, 139–141
evolution of, 140–141
function of, 139
gestural, as precursor, 140
of psychology, 158–159
Learning, ubiquity of, 19
of values, 87
Liberty, conditions of, 135–136
Limited capacity, of consciousness, 47, 52–54
and emotion, 73
of memory, 17
Lust, contrasted with other emotions, 75

Materialism and psychology, 152–153
Mind, definitions of, 40–41
and consciousness, 41–42, 48

Mind-Body relationship, 42–43, 143–144
and reductionism, 143
Modelling of organisms, 13–15, 41
Morality, 123–127
and anti-semitism, 125–126
and moral behavior in primates, 125
and rationality, 126–127
religions as source of, 123–124
universality and absolutes in, 126–127
Myths and gods, 10–11

Novelty, defined, 95
and emotion, 96

Object constancy, 18–19

Peacemaking, in primates, 106
Phobias, 17
Physics, as model for psychology, 166
Positive and negative freedoms, *see* freedom
Postmodernism and psychology, 165
Power, psychology of, 133–135
definition of, 133–134
and group identification, 135
Pre-adaptation, *see* exaptation
Preferences, bases of, 92–93

Reductionism, *see* Mind-Body relationship

Schemas, 43–45, 49, 57
definition of, 44
and incongruity, 44
of values, 87
Science, ancient, 11–12
as cultural product, 148–149
modern, 13–14
Selfish gene, defined, 23–24
Sex differences, 118–121
cognitive, 119
social roles, 119–120
and child rearing, 120
Sexual behavior, variations of, 118–119
early church views of, 124
Society, functions of, 100–101
parallel developments in, 101
Sociobiology, 32–34
and matrilineal culture, 121

Solutions, evolutionary, *see* evolution, parallel
Structure, and representation, 141–143
Sympathetic nervous system, 70–71, 73
  and its role in aggression, 107
  as source of "hot" emotions, 73

Taste, *see* value
Theory of emotion,
  Frijda's, 79
  Izard's, 79
  James', 67
  Lazarus', 80
  Oatley and Johnson-Laird's, 80
  Ortony, Clore & Collins', 79
  Schachter's, 69
Thought, organization of, 138–139

Values, 85–99
  beauty as, 89
  definitions of, 86, 88–89
  and emotion, 69, 89
  and schemas, 91–92
  social contexts of, 97–98
  sources of, 90
  tastes as, 86

Worldviews, constrained and unconstrained, 5–6
Wundt, 149–156
  and German politics, 153
  his life, 149–151
  and *Völkerpsychologie,* 154–155